Barron's Review Course Series

Let's Review:

Sequential Mathematics, Course III

Third Edition

Lawrence S. Leff

Assistant Principal, Mathematics Supervision
Franklin D. Roosevelt High School
Brooklyn, New York

BARRON'S

— TO RHONA ——————————————————

*For the gifts of
love and friendship.*

All inquiries should be addressed to:
Barron's Educational Series, Inc.
250 Wireless Boulevard
Hauppauge, New York 11788
http://www.barronseduc.com

Library of Congress Catalog Card No. 2002071747

International Standard Book No. 0-7641-2091-3

Library of Congress Cataloging-in-Publication Data

Leff, Lawrence S.
    Let's review. Sequential mathematics, course III/Lawrence S. Leff.–3rd ed.
        p. cm. – (Barron's review course series)
    Includes index.
    ISBN 0-7641-2091-3
    1. Mathematics.   2. Mathematics–Study and teaching (Secondary)—
New York (State)   I. Title: Sequential mathematics, course III.   II. Title:
Sequential mathematics, course 3.   III. Title: Sequential mathematics, course
three.   IV. Title.   V. Series.

QA39.3 .L443 2002
510–dc21   2002071747

PRINTED IN THE UNITED STATES OF AMERICA

9 8 7 6 5 4 3 2 1

TABLE OF CONTENTS

PREFACE

For which course can this book be used?

This new edition reflects the latest trends in the Regents Examination for Course III of the New York State Three-Year Sequence in High School Mathematics. Since this book offers complete topic coverage as well as systematic Regents Exam preparation for Course III, it can be used either as a supplementary study aid or as the main textbook in classes that follow the Course III curriculum.

What special features does this book have?
- *Discussion of Scientific and Graphing Calculators*
  There is a special introductory section on what you need to know about scientific calculators. Throughout the text, the use of scientific calculators is explained whenever appropriate, and typical calculator-key-pressing sequences are displayed. Graphing-calculator-based solutions are explained in the appendix entitled "What Is the Mathematics B Regents Examination?" at the back of the book.
- *A Compact Format Designed for Self-Study and Rapid Learning*
  For easy reference, the major topics of the book are grouped by the appropriate branches of mathematics. The friendly, straightforward writing style quickly identifies essential ideas while avoiding unnecessary details.
- *Key Ideas*
  Each section of each chapter begins with KEY IDEAS that highlight and motivate the material that follows. The KEY IDEAS, together with the helpful diagrams and numerous step-by-step demonstration examples, aim to anticipate and answer students' "why" questions.
- *Regents Examination Preparation*
  The demonstration examples throughout and the practice exercises at the end of each section use Regents Exam types of questions. Each chapter closes with a set of REGENTS TUNE-UP exercises selected from actual past Course III Regents Examinations. These Regents problems provide a comprehensive review of the material covered in the chapter while previewing the types and the levels of difficulty of problems found on actual Regents Examinations. To provide a culminating activity, several full-length Course III Regents Examinations with answers appear at the end of the book.
- *Answers to Hundreds of Practice Exercises*
  The answers to most of the review exercises are given at the end of each chapter. In addition, the answers to *all* of the Regents Tune-up exercises are provided.
- *Glossary of Course III Terms*
  Important Course III mathematical terms are collected for easy reference in a glossary at the end of the book.

Who should use this book?

Students who are having difficulty keeping up with their classroom lessons or who want to advance on their own will find this book well suited to their special needs. Students who want to sharpen their understanding and improve their test grades will benefit greatly from the easily understood explanations, helpful examples, and past Regents Exam questions organized by topic. Students preparing for the Course III Regents Examination will find this book particularly useful whenever they need further explanation or more practice on a troublesome topic.

Since this new edition is designed to be compatible with all styles of classroom teaching and curriculum organization, classroom *teachers* will want to include the book in their personal and department libraries. Teachers will find that it is an ideal companion to any of the existing Course III textbooks, providing a valuable lesson-planning aid as well as a rich source of classroom exercises, homework problems, and test questions.

LAWRENCE S. LEFF
April 2002

SCIENTIFIC CALCULATORS: WHAT YOU NEED TO KNOW

All students are required to have access to scientific calculators when they take the New York State Course III Regents Examination. Graphing calculators are not allowed for this examination. This book assumes that you already have a scientific calculator and are familiar with its main features. Since logarithmic and trigonometric tables will not be provided in Regents Examination booklets, you will need to kow how to use your scientific calculator to do the following:

- Find the values of logarithms of positive numbers.
- Find the antilogarithms of numbers to a stated level of accuracy.
- Find the values of trigonometric functions of angles expressed in degrees, degrees and minutes, and radians.
- Find the degree or radian measures of angles using inverse trigonometric functions. Answers may be required to the nearest degree, nearest minute, nearest 10 minutes, or nearest tenth or hundredth of a degree or radian.

Notation and Conventions

In this book, all calculator keys, except those for numbers, will be boxed. For example, the key-pressing sequence

$$16 \boxed{\div} 2 \boxed{=}$$

means enter 16, press the calculator key labeled with the division symbol, enter 2, and then press the calculator key labeled with the equals symbol.

To access the special functions, operations, or values that are printed directly *above* certain calculator keys, you must first press the $\boxed{\text{2nd}}$, $\boxed{\text{SHIFT}}$, or $\boxed{\text{INV}}$ key, depending on your particular calculator. To simplify matters, this book will always refer to this special key as the $\boxed{\text{2nd}}$ function key.

Optional Calculator Features You Need to Know

In taking the Regents Examination, you are permitted to use all of the special features of your scientific calculator. Therefore, it will be to your advantage to know how to do these things:

- Evaluate numbers with fractional exponents using the "x to y" $\boxed{x^y}$ function key. Some calculators label this key $\boxed{y^x}$.
- Evaluate expressions using keys that calculate factorials $x!$, permutations ${}_nP_r$, and combinations ${}_nC_r$ keys.

- Calculate the mean x and the standard deviation σ_n of a set of n data values for a stated level of accuracy.

Cautions

Unfortunately, all scientific calculators do not work in the same way, nor is the arrangement or labeling of the keys the same. If any calculator procedure described in this book does not work with your particular calculator, you will need to read the instruction manual that came with the calculator.

CALCULATOR TIPS

- Press the $\boxed{\text{AC}}$ or *A*ll *C*lear key before performing a new calculation.
- Make sure the angular mode of the calculator is set to the correct unit of angle measurement. For example, if you are working with angles measured in degrees but RADians appears in the display window, you will need to change the angular mode of the calculator from radians to degrees.
- Remember that, when performing a chain calculation involving a sequence of arithmetic operations, or when applying a special calculator function to an arithmetic computation, you may need to use parentheses. If in doubt, use parentheses. For example, to find the value of $8^{\frac{1}{3}}$ using the $\boxed{x^y}$ function key, let $x = 8$ and $y = (1 \div 3)$. Verify that the key sequence

$$8 \;\boxed{x^y}\; \boxed{(}\; 1 \;\boxed{\div}\; 3 \;\boxed{)}\; \boxed{=}$$

produces 2 in the display window. Thus, $8^{\frac{1}{3}} = 2$.
- First estimate an answer, then use your calculator to obtain the actual answer. If your calculator result varies greatly from the estimate, start over. You may have inadvertently pressed the wrong key.
- Reduce the chances of introducing round-off errors by keeping the results of intermediate calculations in your calculator. For example, to find the value of

$$\frac{19.30 \times \sin 50°}{34.60}$$

correct to the *nearest hundredth*, perform one long calculation using this key sequence:

$$19.3 \;\boxed{\times}\; 50 \;\boxed{\text{SIN}}\; \boxed{\div}\; 34.6 \;\boxed{=}\; .$$

If this keystroke sequence does not work with your calculator, press the $\boxed{\text{SIN}}$ key before entering 50. Since the display of a typical calculator shows 0.427302247, the answer is 0.43, correct to the *nearest hundredth*.

Unit
One

ALGEBRAIC METHODS AND COMPLEX NUMBERS

CHAPTER 1

REVIEW AND EXTENSION OF ALGEBRAIC METHODS

1.1 DEFINING ZERO, NEGATIVE INTEGER, AND FRACTIONAL EXPONENTS

KEY IDEAS

A positive **exponent** represents the number of times a quantity, called the **base**, appears as a factor. Thus, the exponent *4* in the expression 3^4 indicates that the base *3* is to be used as a factor 4 times, as in

$$3^4 = 3 \cdot 3 \cdot 3 \cdot 3.$$

The laws that apply to quantities raised to positive integer powers hold true also for zero, negative integer, and fractional exponents.

Zero and Negative Integer Exponents

Any nonzero quantity raised to the zero power is 1.

Examples: *1.* $5^0 = \mathbf{1}$ *2.* $(-5)^0 = \mathbf{1}$

Any nonzero quantity raised to a negative integer power is equal to the reciprocal of the same quantity with the exponent changed from negative to positive.

Examples: *1.* $x^{-3} = \dfrac{\mathbf{1}}{\boldsymbol{x^3}}$ $(x \neq 0)$

2. $\dfrac{1}{y^{-5}} = \boldsymbol{y^5}$ $(y \neq 0)$

The Laws of Exponents summarized in Table 1.1 apply to all integer-valued exponents.

1

TABLE 1.1 LAWS OF INTEGER EXPONENTS

Law	Rule	Example
Product	$x^a \cdot x^b = x^{a+b}$	$x^3 \cdot x^2 = x^{2+3} = x^5$
Quotient	$\dfrac{x^a}{x^b} = x^{a-b}$	$\dfrac{x^3}{x^2} = x^{3-2} = x^1 = x$
Power	$\left(x^a\right)^b = x^{ab}$	$\left(x^3\right)^2 = x^{3 \cdot 2} = x^6$
Power of Product	$(xy)^a = x^a y^a$	$(xy)^3 = x^3 y^3$
Power of Quotient	$\left(\dfrac{x}{y}\right)^a = \dfrac{x^a}{y^a}$	$\left(\dfrac{x}{y}\right)^2 = \dfrac{x^2}{y^2} \quad (y \neq 0)$

A fraction raised to a negative integer power can be simplified by inverting the fraction and changing the sign of the exponent from negative to positive, as in

$$\left(\frac{5}{2}\right)^{-3} = \left(\frac{2}{5}\right)^3 = \frac{2^3}{5^3} = \frac{8}{125}.$$

===== **MATH FACTS** =====

ZERO AND NEGATIVE INTEGER EXPONENT RULES

- $x^0 = 1 \quad (x \neq 0)$

- $x^{-n} = \dfrac{1}{x^n}$ and $x^n = \dfrac{1}{x^{-n}} \quad (x \neq 0)$

- $\left(\dfrac{a}{b}\right)^{-n} = \left(\dfrac{b}{a}\right)^n \quad (a, b \neq 0)$

Examples

1. Rewrite in simplest form, using a positive exponent:

(a) $2x^{-4}$ (b) $\dfrac{1}{2x^{-4}}$ (c) $(2x)^{-4}$

Solutions:

(a) $2x^{-4} = \dfrac{2}{x^4}$

(b) $\dfrac{1}{2x^{-4}} = \dfrac{x^4}{2}$

(c) $(2x)^{-4} = \dfrac{1}{(2x)^4}$
$= \dfrac{1}{2^4 x^4} = \dfrac{1}{16x^4}$

2. Rewrite in simplest form, using positive exponents:

(a) $\dfrac{3x^{-2}}{5y^{-3}}$ **(b)** $\left(\dfrac{r^3}{s^2}\right)^{-4}$ **(c)** $\dfrac{12x^{-5}y^7}{6y^3}$

Solutions:

(a) $\dfrac{3x^{-2}}{5y^{-3}} = \dfrac{3y^3}{5x^2}$

(b) $\left(\dfrac{r^3}{s^2}\right)^{-4} = \left(\dfrac{s^2}{r^3}\right)^4 = \dfrac{s^8}{r^{12}}$

(c) $\dfrac{12x^{-5}y^7}{6y^3} = \dfrac{12y^{7-3}}{6x^5}$

$\qquad\qquad = \dfrac{2y^4}{x^5}$

3. If $x = 3$, find the value of each expression:

(a) $2x^0 + x^{-2}$ **(b)** $(2x)^0 - 6x^{-2}$

Solutions: Replace x by 3 and simplify:

(a) $2x^0 + x^{-2} = 2(3^0) + 3^{-2}$

$\qquad\qquad = 2 \cdot 1 + \dfrac{1}{3^2}$

$\qquad\qquad = 2 + \dfrac{1}{9}$

$\qquad\qquad = 2\dfrac{1}{9}$ or $\dfrac{19}{9}$

(b) $(2x)^0 - 6x^{-2} = (2 \cdot 3)^0 - 6(3)^{-2}$

$\qquad\qquad = 6^0 - \dfrac{6}{3^2}$

$\qquad\qquad = 1 - \dfrac{6}{9}$

$\qquad\qquad = \dfrac{3}{9}$ or $\dfrac{1}{3}$

Square Root

A **square root** of a nonnegative number is one of two identical factors whose product is the number. For example, a square root of 16 is 4 since $4^2 = 16$. Thus, $\sqrt{16} = 4$, where the symbol $\sqrt{}$ is called a **radical sign** and 16, the number underneath the radical sign, is the **radicand**. The expression $\sqrt{16}$ is read as "the square root of 16" or as "radical 16."

Every positive number has two square roots. For example, the two square roots of 16 are 4 and –4 since $4^2 = 16$ and $(-4)^2 = 16$. The positive square root is called the *principal* square root. If a number has more than one real root, the root that has the same sign as the original number is called the **principal root**. The expression $\sqrt{x}$ always names the *principal* square root of x. Thus, $\sqrt{16} = 4$ rather than $\sqrt{16} = -4$.

Not all real numbers have square roots. For example, $\sqrt{-9}$ does not name a real number since it is not possible to find two identical real numbers whose product is negative 9.

The *k*th Root of a Number

The **cube root** of a number is one of three identical factors whose product is the number. The principal cube root of 8 is 2 since $2^3 = 8$. The principal cube root of -8 is -2 since $(-2)^3 = -8$. Thus

$$\sqrt[3]{8} = 2 \quad \text{and} \quad \sqrt[3]{-8} = -2.$$

A **fourth root** of a nonnegative number is one of four identical factors whose product is the number. There are two real fourth roots of 81, 3 and -3, since $3^4 = 81$ and $(-3)^4 = 81$. The principal fourth root, however, is 3 since it has the same sign as 81. Thus

$$\sqrt[4]{81} = 3.$$

In general:

- The number k in $\sqrt[k]{x}$ is called the **index** of the radical and indicates what root is to be taken. Therefore, k represents some integer greater than or equal to 2. If the index does not appear, as in square root radicals, it is assumed to be 2. Thus, $\sqrt[2]{9}$ and $\sqrt{9}$ mean the same thing.

- If the index k is even and $x < 0$, then $\sqrt[k]{x}$ is not real.

Using Exponents to Indicate Roots

By extending Table 1.1 to include fractional exponents, the Product Law of Exponents may be used to give meaning to these exponents. Since

$$x^{\frac{1}{2}} \cdot x^{\frac{1}{2}} = x^{\frac{1}{2}+\frac{1}{2}} = x^1,$$

$x^{\frac{1}{2}}$ represents one of two identical factors whose product is x. Thus, $x^{\frac{1}{2}} = \sqrt[2]{x}$ or $\sqrt{x}$. Similarly, $x^{\frac{1}{3}} = \sqrt[3]{x}$. In general, $x^{\frac{1}{k}}$ represents the kth root of x.

MATH FACTS

MEANING OF $x^{\frac{1}{k}}$

$$x^{\frac{1}{k}} = \sqrt[k]{x}$$

where k is an integer ≥ 2 and $x \geq 0$ whenever k is even.

Interpreting Fractional Exponents

The quantity $x^{\frac{n}{k}}$, where $k \neq 0$ and $\frac{n}{k}$ is in lowest terms, can be evaluated in two different ways:

- Since $x^{\frac{n}{k}} = \left(x^{\frac{1}{k}}\right)^n$, the quantity $x^{\frac{n}{k}}$ may be evaluated by finding the kth root of the base x, and then raising the result to the nth power. For example,

$$27^{\frac{2}{3}} = \left(\sqrt[3]{27}\right)^2 = (3)^2 = \mathbf{9}.$$

- Since it is also true that $x^{\frac{n}{k}} = \left(x^n\right)^{\frac{1}{k}}$, the quantity $x^{\frac{n}{k}}$ may be evaluated by raising the base x to the nth power, and then finding the kth root of the result. For example,

$$27^{\frac{2}{3}} = \sqrt[3]{27^2} = \sqrt[3]{729} = \mathbf{9}.$$

=========================== **MATH FACTS** ===========================

MEANING OF $x^{\frac{n}{k}}$

$$x^{\frac{n}{k}} = \sqrt[k]{x^n} = \left(\sqrt[k]{x}\right)^n$$

where k is an integer ≥ 2 and $x \geq 0$ whenever k is even.

Examples

4. Rewrite with a fractional exponent and simplify:

(a) $\sqrt{y^8}$ **(b)** $\sqrt[3]{8x^6}$

Solutions: **(a)** $\sqrt{y^8} = \left(y^8\right)^{\frac{1}{2}} = y^{\frac{8}{2}} = \mathbf{y^4}$ | **(b)** $\sqrt[3]{8x^6} = \left(8x^6\right)^{\frac{1}{3}}$

$$= (8)^{\frac{1}{3}}\left(x^6\right)^{\frac{1}{3}}$$

$$= 2 \cdot x^{\frac{6}{3}} = \mathbf{2x^2}$$

5. If $x = 16$, find the value of each expression:

(a) $2x^{\frac{1}{2}} - x^{\frac{3}{4}}$ **(b)** $\left(\dfrac{2}{x}\right)^{-\frac{3}{2}}$

Solutions: Replace x by 16 and simplify:

(a) $2x^{\frac{1}{2}} - x^{\frac{3}{4}} = 2(16)^{\frac{1}{2}} - 16^{\frac{3}{4}}$

$\qquad = 2\sqrt{16} - \left(\sqrt[4]{16}\right)^3$

$\qquad = 2 \cdot 4 \quad - (2)^3$

$\qquad = 8 \qquad - 8$

$\qquad = 0$

(b) $\left(\dfrac{2}{x}\right)^{-\frac{2}{3}} = \left(\dfrac{2}{16}\right)^{-\frac{2}{3}} = \left(\dfrac{16}{2}\right)^{\frac{2}{3}}$

$\qquad = 8^{\frac{2}{3}}$

$\qquad = \left(\sqrt[3]{8}\right)^2$

$\qquad = (2)^2$

$\qquad = 4$

Using a Scientific Calculator to Evaluate $x^{\frac{p}{k}}$

The $\boxed{x^y}$ key can be used to evaluate a number raised to a fractional power. For example, the value of $8^{-\frac{2}{3}}$ is found by keying in

$$8 \;\boxed{x^y}\; \boxed{(} \; 2 \; \boxed{\pm} \; \boxed{\div} \; 3 \; \boxed{)} \; \boxed{=} \;.$$

Since the display window shows 0.25, $8^{-\frac{2}{3}} = 0.25$. If your calculator does not have a "x to the y" function key, look for a "y to the x" function key.

Exercise Set 1.1

1–6. Rewrite in simplest form, using a positive exponent.

1. $3y^{-2}$ **3.** $\dfrac{1}{3y^{-2}}$ **5.** $\left(x^{-2}y^{-3}\right)^2$

2. $\dfrac{(3y)^{-2}}{3}$ **4.** $\dfrac{2x^{-3}}{3y^{-2}}$ **6.** $\left(\dfrac{a^{-3}}{b^{-2}}\right)^{-2}$

7. If $10^x = m$, express each of the following in terms of m:
(a) 10^{x+3} (b) 10^{3x} (c) 10^{x-2} (d) $10^{\frac{x}{2}}$

8–11. Rewrite in simplest form, using positive exponents.

8. $\dfrac{15x^{-2}}{10x^4y^{-3}}$ **9.** $\left(\dfrac{p^3}{q^2}\right)^{-3}$ **10.** $\left(\dfrac{u^{-2}}{v^{-3}}\right)^{-1}$ **11.** $\dfrac{16a^5b^{-1}}{24a^{-3}b^2}$

12–19. If $x = 4$, find the value of each expression.

12. $2x^0 + x^{-2}$ **14.** $x^{-\frac{1}{2}}$ **16.** $(2x)^{\frac{1}{3}}$ **18.** $\left(x^{-3}\right)^{\frac{1}{2}}$

13. $2(x+1)^0 - 6x^{-2}$ **15.** $x^{\frac{3}{2}}$ **17.** $(16x)^{\frac{2}{3}}$ **19.** $\left(\dfrac{x}{3}\right)^{-2}$

20–23. Evaluate.

20. $64^{\frac{2}{3}}$ **21.** $64^{-\frac{3}{2}}$ **22.** $\left(\dfrac{25}{16}\right)^{\frac{3}{2}}$ **23.** $\left(\dfrac{25}{16}\right)^{-\frac{3}{2}}$

24–26. If x = 8, find the value of each expression.

24. $x^{\frac{2}{3}} - (x+1)^{\frac{3}{2}}$ **25.** $x^{-\frac{4}{3}} + (3x)^{0}$ **26.** $\dfrac{1}{x^2} + x^{-2}$

27–31. Write in exponential form, and then simplify.

27. $\sqrt{y^{10}}$ **28.** $\sqrt{a^2 b^6}$ **29.** $\sqrt[3]{27x^{12}}$ **30.** $\sqrt{4m^2 n}$ **31.** $\sqrt[3]{-8x^6 y^2}$

1.2 FACTORING ALGEBRAIC EXPRESSIONS

KEY IDEAS

The polynomial $x^2 + 5x$ is factored when it is written as $x(x + 5)$. The polynomials x and $x+5$ are factors of the original polynomial. **Factoring** a polynomial means writing it as the product of two or more other polynomials.

To verify that $x^2 + 5x$ can be factored as $x(x + 5)$, multiply each term inside the parentheses by x. The result should be the original expression:

$$x(x + 5) = x \cdot x + x \cdot 5 = x^2 + 5x.$$

Factoring can be used in simplifying algebraic fractions and in solving quadratic equations.

Factoring Polynomials

Some polynomials can be factored by recognizing that each term of the polynomial contains a common numerical factor, a common literal factor, or both.

Examples: *1.* $2a - 8b = \mathbf{2(a - 4b)}$
2. $10y^2 + 15y = \mathbf{5y(2y + 3)}$
3. $6r^2 s - 9rs^2 + 12r^2 s^2 = \mathbf{3rs(2r - 3s + 4rs)}$
4. $-3x - 7y = \mathbf{-(3x + 7y)}$

Unit One ALGEBRAIC METHODS AND COMPLEX NUMBERS

Factoring the Difference of Two Squares

The binomials $p + q$ and $p - q$ are called *conjugate* binomials since they take the sum and difference of the *same* two terms. Any binomial of the form $p^2 - q^2$ can be factored as the product of the conjugate binomials $p + q$ and $p - q$.

Examples: 1. $y^2 - 36 = (y + 6)(y - 6)$
2. $4a^2 - 25b^2 = (2a + 5b)(2a - 5b)$

MATH FACTS

FACTORING THE DIFFERENCE OF TWO SQUARES

$$p^2 - q^2 = (p + q)(p - q)$$

Factoring $x^2 + bx + c$

Two binomials can be multiplied horizontally by using the familiar FOIL method, where F represents the product of the <u>F</u>irst terms of the binomials, O and I represent the products of the <u>O</u>utermost and <u>I</u>nnermost terms of the binomials, and L represents the product of the <u>L</u>ast terms of the binomials.

$$\overset{\textbf{F}\quad\quad\textbf{O}\quad\ \ \textbf{I}\quad\quad\textbf{L}}{}$$

Example: $(x + 5)(x - 2) = x^2 + (-2x + 5x) + [(5)(-2)]$
$= x^2 + 3x - 10$

Using the reverse of this process, begin factoring $x^2 + bx + c$ by writing

$$x^2 + bx + c = (x + \underline{\ \ })(x + \underline{\ \ }).$$

The blanks must be filled with two numbers whose product is c *and* whose sum is b. For example, start factoring $x^2 - 7x - 18$ by writing

$$x^2 - 7x - 18 = (x + \underline{\ \ })(x + \underline{\ \ }).$$

Next, identify the pairs of factors of -18. Although -18 has four pairs of factors $(1, -18; -1, 18; 9, -2; -9, 2)$, only -9 and 2 have a sum of -7. Therefore fill in the blanks with -9 and 2. Hence

$$x^2 - 7x - 18 = (x - 9)(x + 2).$$

Factoring $ax^2 + bx + c\ (a \neq 1)$

When determining the possible binomial factors of $ax^2 + bx + c$, where a is different from 1, the factors of a must be taken into consideration. For example, the binomial factors of $4x^2 - x - 5$ take the form

$$(\Box x + \underline{\quad})(\Box x + \underline{\quad}).$$

Since 4 has two pairs of positive factors (2, 2 and 4, 1), more than one pair of numbers can be entered in the boxes as coefficients of x:

$$4x^2 - x - 5 = (2x + \underline{\quad})(2x + \underline{\quad}) \text{ or } 4x^2 - x - 5 = (4x + \underline{\quad})(x + \underline{\quad}).$$

Next, identify the pairs of factors of $-5(-5, 1$ and $5, -1)$. Each pair of factors must be tested by placing the factors in the blanks in each pair of possible binomial factors. By trial and error, you determine that the correct pair of factors is -5 and 1; position them in the binomials in such a way that the sum of the outermost and innermost products is $-x$:

$$4x^2 - x - 5 = (\mathbf{4x - 5})(\mathbf{x + 1}).$$

To check that this is the correct factorization, use FOIL to verify that the product $(4x - 5)(x + 1)$ equals $4x^2 - x - 5$.

Factoring $ax^2 + bx + c$ by Decomposing bx

If a quadratic trinomial of the form $ax^2 + bx + c$ can be factored, its binomial factors can be identified by rewriting the middle term, bx, as the sum of two other terms and then factoring. The numerical coefficients of these two terms must be chosen so that their sum is b and their product is $a \cdot c$. To factor $3x^2 + 10x + 8$ using this method the procedure is as follows:

- Find two integers whose sum is $+10$ (the coefficient of x), and whose product is 24 (the product of 3 and 8). Since these integers are 4 and 6, write:

$$3x^2 + 10x + 8 = 3x^2 + \underbrace{4x + 6x}_{+10x} + 8$$

- Group the first and last pairs of terms:

$$= \left(3x^2 + 4x\right) + \left(6x + 8\right)$$

- Factor out the greatest common factor of each pair of terms:

$$= x(3x + 4) + 2(3x + 4)$$

- Factor out the common *binomial* factors:

$$= (3x + 4)(x + 2)$$

Thus, $3x^2 + 10x + 8 = (\mathbf{3x + 4})(\mathbf{x + 2})$.

Factoring Completely

A polynomial is factored *completely* when each of its factors cannot be factored further. To factor completely, it may be necessary to apply more than one factoring technique. Typically, begin by removing a common monomial factor, if possible, and then factoring further, again if possible.

Examples: 1. $3x^3 - 12x = 3x(x^2 - 4)$
$$= 3x(x+2)(x-2)$$
2. $t^3 + 6t^2 - 16t = t(t^2 + 6t - 16)$
$$= t(t+8)(t-2)$$

Simplifying Fractions

If the numerator and the denominator of a fraction have a factor in common, the fraction can be simplified by factoring the numerator and the denominator and then canceling pair(s) of common factors in the numerator and denominator since their quotient is 1. This technique is used in simplifying arithmetic as well as algebraic fractions.

Examples: 1. $\dfrac{x^2 + 3x}{x^2 - 9} = \dfrac{x\cancel{(x+3)}}{(x-3)\cancel{(x+3)}} = \dfrac{x}{x-3}$

2. $\dfrac{s-t}{2t-2s} = \dfrac{s-t}{2(t-s)} = \dfrac{-\cancel{(t-s)}}{2\cancel{(t-s)}} = -\dfrac{1}{2}$

Avoiding a Zero Denominator

Since division by 0 is not defined, the variable or variables that appear in an algebraic fraction are assumed to be restricted in value so that the denominator can never be equal to 0. To identify the particular value or values of the variable that make an algebraic fraction undefined, rewrite the denominator and set it equal to 0. The root or roots of this equation represent the value or values that the variable *cannot* equal.

For example, to find the value or values of x that make the fraction

$$\frac{x^2 + 1}{x^2 - x - 12}$$

undefined, set $x^2 - x - 12$ equal to 0, and solve the resulting equation:

$$x^2 - x - 12 = 0$$

Factor: $$(x+3)(x-4) = 0$$

Apply the Zero Product Rule: $$x + 3 = 0 \text{ or } x - 4 = 0$$

Solve each equation: $$x = -3 \text{ or } x = 4$$

The fraction is undefined when $x = -3$ or $x = 4$.

Exercise Set 1.2

1–9. Factor if possible.

1. $24x^3y + 30x^2y^5$ **4.** $6a^3 - 9a^2b$ **7.** $8u^5w^2 - 40u^2w^5$

2. $20a^2c - 32a^3b^2$ **5.** $x^3 - x^2 + x$ **8.** $p^2k^4 - p^3k^2 + (pk)^2$

3. $21 + 2w$ **6.** $-ab - b$ **9.** $(x+a)^2 - a(x+a)$

10–18. Factor as the product of two binomials.

10. $a^2 - 100$ **13.** $t^2 + 4t - 21$ **16.** $3x^2 - 2x - 21$

11. $\dfrac{x^2}{4} - 36$ **14.** $x^2 + 2xy + y^2$ **17.** $5s^2 + 14s - 3$

12. $0.16 - b^2$ **15.** $2q^2 - q - 15$ **18.** $4x^2 + 3x - 7$

19–27. Factor completely.

19. $2y^2 - 50$ **22.** $10y^3 + 50y^2 - 500y$ **25.** $2p^3 + p^2 - 15p$

20. $81 - b^4$ **23.** $-5t^2 + 5$ **26.** $x^2 - (x + 42)$

21. $45 - 4x - x^2$ **24.** $x^4 - y^4$ **27.** $c(c-4) - 45$

28 and 29. Cross-multiply, and then solve for x.

28. $\dfrac{x-4}{2} = \dfrac{3x}{x+4}$ **29.** $\dfrac{x-2}{x} = \dfrac{x+4}{3x}$

30–35. Write in lowest terms.

30. $\dfrac{-x-y}{x+y}$ **32.** $\dfrac{3x+6}{x^2-4}$ **34.** $\dfrac{2x^2-50}{2x^2+14x+20}$

31. $\dfrac{3x+15}{x^2+5x}$ **33.** $\dfrac{b-a}{a^2-b^2}$ **35.** $\dfrac{x^2-y^2}{x^2-2xy+y^2}$

36–38. State for which value(s) of x *the fraction is undefined.*

36. $\dfrac{3}{x^2+x-2}$ **37.** $\dfrac{x+2}{x^2-2x}$ **38.** $\dfrac{x+5}{5^x-1}$

11

1.3 MULTIPLYING AND DIVIDING ALGEBRAIC FRACTIONS

⌃ KEY IDEAS ⌃

Arithmetic fractions are multiplied by writing the product of their numerators over the product of their denominators. To have the product appear in simplest form, it is often easier to cancel common pairs of factors in the numerator and denominator *before* multiplying:

$$\frac{4}{9} \cdot \frac{3}{10} = \frac{\overset{2}{\cancel{4}}}{\underset{3}{\cancel{9}}} \cdot \frac{\overset{1}{\cancel{3}}}{\underset{5}{\cancel{10}}} = \frac{2}{15}.$$

Algebraic fractions are multiplied in much the same way.

To divide one fraction by another, invert the second fraction and then multiply.

Multiplying Algebraic Fractions

To multiply algebraic fractions, begin by factoring the numerator and denominator of each fraction, if possible. Then cancel common factors whose quotient is 1.

Example

1. Write the product in the simplest form:

$$\frac{x^2 - 9}{10} \cdot \frac{15}{(x+3)^2}.$$

Solution: Begin by factoring the numerators and denominators:

$$\frac{x^2 - 9}{10} \cdot \frac{15}{(x+3)^2} = \frac{(x-3)(x+3)}{2 \cdot 5} \cdot \frac{5 \cdot 3}{(x+3)(x+3)}$$

Cancel common factors whose quotient is 1:

$$= \frac{(x-3)\overset{1}{\cancel{(x+3)}}}{2 \cdot \cancel{5}} \cdot \frac{\overset{1}{\cancel{5}} \cdot 3}{\cancel{(x+3)}(x+3)}$$

Multiply the remaining factors:

$$= \frac{3(x-3)}{2(x+3)}$$

Dividing Algebraic Fractions

To divide algebraic fractions, change the example into an equivalent multiplication problem by inverting the second fraction. Example 2 illustrates that you may need to use a variety of factoring techniques to identify common pairs of factors in the numerators and denominators of the fractions.

Example

2. Express each quotient in simplest form:

(a) $\dfrac{6x - 3x^2}{9x^3 - x} \div \dfrac{x^2 - 4}{3x^2 + 5x - 2}$ (b) $\dfrac{x^2 + 2xy + y^2}{xy - x} \div \dfrac{x^2 - y^2}{y^2 - y}$

Solutions: (a) Invert the second fraction and multiply:

$$\frac{6x - 3x^2}{9x^3 - x} \div \frac{x^2 - 4}{3x^2 + 5x - 2} = \frac{6x - 3x^2}{9x^3 - x} \cdot \frac{3x^2 + 5x - 2}{x^2 - 4}$$

Factor the numerators and denominators completely:

$$= \frac{3x(2 - x)}{x(3x + 1)(3x - 1)} \cdot \frac{(3x - 1)(x + 2)}{(x - 2)(x + 2)}$$

Cancel common factors whose quotient is 1:

$$= \frac{3\cancel{x}(2 - x)}{\cancel{x}(3x + 1)\cancel{(3x - 1)}} \cdot \frac{\overset{1}{\cancel{(3x - 1)}}\cancel{(x + 2)}}{(x - 2)\cancel{(x + 2)}}$$

Since $(2 - x) = -(x - 2)$, cancel $(2 - x)$ over $(x - 2)$ as -1:

$$= \frac{3\overset{-1}{\cancel{(2 - x)}}}{3x + 1} \cdot \frac{1}{\cancel{(x - 2)}}$$

Multiply the remaining factors:

$$= \frac{-3}{3x + 1}$$

(b) $\dfrac{x^2 + 2xy + y^2}{xy - x} \div \dfrac{x^2 - y^2}{y^2 - y} = \dfrac{\overset{1}{\cancel{(x + y)^2}}}{x\cancel{(y - 1)}} \cdot \dfrac{y\overset{1}{\cancel{(y - 1)}}}{\cancel{(x + y)}(x - y)}$

$$= \frac{y(x + y)}{x(x - y)}$$

Exercise Set 1.3

1–14. Express each product or quotient in simplest form.

1. $\dfrac{12y^2}{x^2+7x} \cdot \dfrac{x^2-49}{2y^5}$

8. $\dfrac{x^2-3x}{x^2+3x-10} \div \dfrac{x^2-x-6}{x^2-4}$

2. $\dfrac{8m^3}{3} \div \dfrac{6m^3}{3m-12}$

9. $\dfrac{3p+6}{p^2-9} \div \dfrac{p^2+4p+4}{p^2+3p}$

3. $\dfrac{x^2-2x-8}{x^2-25} \cdot \dfrac{2x+10}{x^2-4}$

10. $\dfrac{6a^2b}{x^2-x-72} \div \dfrac{2ab^3}{x^2-64}$

4. $\dfrac{18}{x^2-y^2} \div \dfrac{9}{y-x}$

11. $\dfrac{r^2-s^2}{2rs} \cdot \dfrac{6r^2}{3s-3r}$

5. $\dfrac{(x-7)^2}{x^2-6x-7} \cdot \dfrac{5x+5}{x^2-49}$

12. $\dfrac{x^2+2x}{x^2+2x-15} \cdot \dfrac{2x-6}{4} \div \dfrac{x^2+x-2}{x^2+4x-5}$

6. $\dfrac{a^2-b^2}{2ab} \div \dfrac{a-b}{a^2}$

13. $\dfrac{t^2-1}{t^2-4} \div \dfrac{9t+9}{4t+12} \cdot \dfrac{2-t}{t^2+2t-3}$

7. $\dfrac{x^2-9}{x^2-8x} \cdot \dfrac{x-8}{x^2-6x+9}$

14. $\dfrac{x^2+4xy+3y^2}{x^2-y^2} \cdot \dfrac{x^2+xy}{x-y} \div \dfrac{x^2+3xy}{(x-y)^2}$

1.4 ADDING AND SUBTRACTING ALGEBRAIC FRACTIONS

∧ KEY IDEAS ∕ ⟍

Arithmetic fractions that have the same denominator are combined by writing the sum or difference of their numerators over the common denominator:

$$\frac{2}{7}+\frac{3}{7}=\frac{2+3}{y}=\frac{5}{7}.$$

If the fractions have *different* denominators, each fraction must first be changed into an equivalent fraction having the LCD (Lowest Common Denominator) as its denominator.

Algebraic fractions are handled in much the same way.

Combining Fractions with the Same Denominator

To combine algebraic fractions having the same denominator, write the sum or difference of the numerators over the common denominator and then simplify.

Example 1 illustrates that it is sometimes necessary to remove a factor of -1 from the denominator of a fraction before it can be combined with another fraction.

Example

1. Express as a single fraction in lowest terms: $\dfrac{4x}{x^2-4}+\dfrac{x+6}{4-x^2}$.

Solution: Rewrite $4-x^2$ as $-x^2+4$. Then factor out -1 so that

$$-x^2+4=-\left(x^2-4\right).$$

Rewrite the given expression: $\dfrac{4x}{x^2-4}+\dfrac{x+6}{4-x^2}=\dfrac{4x}{x^2-4}+\dfrac{x+6}{-\left(x^2-4\right)}$

Place the negative sign in front of the second fraction: $=\dfrac{4x}{x^2-4}-\dfrac{x+6}{x^2-4}$

Write the numerators over the common denominator: $=\dfrac{4x-(x+6)}{x^2-4}$

Simplify the numerator: $=\dfrac{4x-x-6}{x^2-4}$

Write the fraction in lowest terms: $=\dfrac{3x-6}{x^2-4}$

Factor, and cancel common factors: $=\dfrac{3\cancel{(x-2)}^{\,1}}{\cancel{(x-2)}(x+2)}$

$=\dfrac{3}{x+2}$

Combining Fractions with Unlike Denominators

To combine fractions having unlike denominators, write the fractions as equivalent fractions having a common denominator by (1) factoring each of the original denominators and then comparing them to find the lowest common denominator (LCD); and (2) multiplying the denominator *and* the numerator of each fraction by the factor or factors needed to change the denominator to the LCD. The fractions can now be added by following the rules for adding fractions having like denominators.

Examples

2. Express the sum in simplest form: $\dfrac{x}{x+1} + \dfrac{x}{x-1}$.

Solution: The Lowest Common Denominator must contain the denominators $x+1$ and $x-1$ as factors. Hence

$$LCD = (x+1)(x-1).$$

Multiply each fraction by 1 in the form of the factor(s) of the LCD that its denominator lacks divided by that factor or factors.

$$\frac{x}{x+1}\left(\frac{x-1}{x-1}\right) + \frac{x}{x-1}\left(\frac{x+1}{x+1}\right) = \frac{x(x-1)+x(x+1)}{(x+1)(x-1)}$$

$$= \frac{(x^2 - x)+(x^2 + x)}{(x+1)(x-1)}$$

$$= \frac{2x^2}{(x+1)(x-1)}$$

3. Express the difference in simplest form: $\dfrac{x}{x^2+3x-4} - \dfrac{x-4}{x^2-1}$.

Solution: Factor each denominator:

$$x^2 + 3x - 4 = (x-1)(x+4)$$
$$x^2 - 1 = (x-1)(x+1)$$
$$LCD = (x-1)(x+4)(x+1)$$

Compare each of the factored denominators to the LCD. The denominator of the first fraction lacks the factor $(x + 1)$, while the second fraction's denominator does not include $(x + 4)$ as a factor. Therefore, multiply each fraction by multiplying the numerator and the denominator of the first fraction by $(x + 1)$, and multiplying the numerator and the denominator of the second fraction by $(x + 4)$.

$$\frac{x}{x^2+3x-4} - \frac{x-4}{x^2-1} = \frac{x}{(x-1)(x+4)} - \frac{x-4}{(x-1)(x+1)}$$

Multiply by a form of 1: $\quad = \dfrac{x}{(x-1)(x+4)}\left(\dfrac{x+1}{x+1}\right) - \dfrac{x-4}{(x-1)(x+1)}\left(\dfrac{x+4}{x+4}\right)$

Combine over the LCD: $\quad = \dfrac{x(x+1)-(x-4)(x+4)}{(x-1)(x+4)(x+1)}$

Simplify the numerator: $= \dfrac{x^2 + x - (x^2 - 16)}{(x-1)(x+4)(x+1)}$

$$= \dfrac{x + 16}{(x-1)(x+4)(x+1)}$$

Exercise Set 1.4

1–16. Write each expression as a single fraction in simplest form.

1. $\dfrac{2x+y}{x+2y} + \dfrac{x+5y}{x+2y}$

9. $\dfrac{a}{a-b} + \dfrac{b}{a+b}$

2. $\dfrac{p}{p^2-9} - \dfrac{3}{p^2-9}$

10. $\dfrac{6}{y^2-9} + \dfrac{4}{(y-3)^2}$

3. $\dfrac{a^2-5}{a-b} + \dfrac{b^2-5}{b-a}$

11. $\dfrac{3}{x^2-4} + \dfrac{2}{x^2+5x+6}$

4. $\dfrac{8x}{3x+12} + \dfrac{2}{2x+8}$

12. $\dfrac{2x}{x^2+7x} - \dfrac{6x-5}{x^2-49}$

5. $\dfrac{5}{b-3} - \dfrac{4}{b}$

13. $\dfrac{5}{r^2-s^2} - \dfrac{3}{(r+s)^2}$

6. $\dfrac{y}{y^2-16} - \dfrac{1}{y+4}$

14. $w - y - \dfrac{1}{w+y}$

7. $\dfrac{7}{x^2-1} + \dfrac{1}{2x+2}$

15. $\dfrac{x+2}{x^2-9} + \dfrac{2}{x^2+x-6}$

8. $\dfrac{a^2+1}{a^2-1} - \dfrac{a}{a+1}$

16. $\dfrac{5}{a^2-3a-10} + \dfrac{3a-2}{a^2-25} - \dfrac{3a}{3a+6}$

17 and 18. Find each product by first simplifying the expressions within the parentheses.

17. $\left(2 + \dfrac{2}{x}\right)\left(\dfrac{1}{x+1} - 1\right)$

18. $\left(\dfrac{1}{x+1} - \dfrac{1}{x-1}\right)^2$

1.5 SIMPLIFYING COMPLEX FRACTIONS

Fractions such as

$$\frac{\frac{4}{5}}{6}, \quad \frac{2}{\frac{9}{10}}, \quad \frac{\frac{2}{7}}{\frac{3}{8}}, \quad \text{and} \quad \frac{\frac{x}{2}+\frac{1}{x}}{\frac{1}{2x}}$$

are examples of *complex fractions*. A **complex fraction** is a fraction that has other fractions in its numerator, in its denominator, or in both its numerator and its denominator. Simplifying a complex fraction means writing it as a "simple" fraction that does not have another fraction in either its numerator or its denominator.

Simplifying a Complex Fraction by Dividing

A complex fraction can be simplified by dividing its numerator by its denominator:

$$\frac{\frac{2}{7}}{\frac{3}{5}} = \frac{2}{7} \div \frac{3}{5} = \frac{2}{7} \cdot \frac{5}{3} = \frac{10}{21}.$$

Example

1. Express in simplest form:

(a) $\dfrac{\dfrac{x}{2}+\dfrac{1}{x}}{\dfrac{1}{2x}}$ (b) $\dfrac{1-x^{-1}}{1-x^{-2}}$

Solutions: **(a)** Rewrite the complex fraction as a division example:

$$\frac{\frac{x}{2}+\frac{1}{x}}{\frac{1}{2x}} = \left[\frac{x}{2}+\frac{1}{x}\right] \div \frac{1}{2x}$$

Invert the last fraction and multiply: $= \left[\dfrac{x}{2} + \dfrac{1}{x} \right] \cdot \dfrac{2x}{1}$

Use the distributive law: $= \left(\dfrac{x}{2} \right) \cdot 2x + \left(\dfrac{1}{x} \right) \cdot 2x$

Simplify: $= x^2 + 2$

(b) Rewrite with positive exponents: $\dfrac{1 - x^{-1}}{1 - x^{-2}} = \dfrac{1 - \dfrac{1}{x}}{1 - \dfrac{1}{x^2}}$

Combine terms in the
numerator and in the
denominator: $= \dfrac{\dfrac{x}{x} - \dfrac{1}{x}}{\dfrac{x^2}{x^2} - \dfrac{1}{x^2}} = \dfrac{\dfrac{x-1}{x}}{\dfrac{x^2-1}{x^2}}$

Divide: $= \left(\dfrac{x-1}{x} \right) \div \left(\dfrac{x^2-1}{x^2} \right)$

Invert the last fraction
and multiply: $= \left(\dfrac{x-1}{x} \right) \cdot \left(\dfrac{x^2}{x^2-1} \right)$

Simplify: $= \dfrac{\cancel{(x-1)}^{1}}{\cancel{x}} \cdot \dfrac{\cancel{x^2}^{x^1}}{\cancel{(x-1)}(x+1)}$

Multiply the remaining factors: $= \dfrac{x}{x+1}$

Simplifying a Complex Fraction by Clearing Fractions

A complex fraction may also be simplified by (1) determining the LCD of
the denominators, and (2) multiplying the complex fraction by 1 in the form
of the LCD divided by itself:

Multiplying by 1

$$\dfrac{\dfrac{2}{7}}{\dfrac{3}{5}} \cdot \left[\dfrac{35}{35} \right] = \dfrac{35 \left(\dfrac{2}{7} \right)}{35 \left(\dfrac{3}{5} \right)} = \dfrac{10}{21}$$

35 is the LCD of 7 and 5.

19

Example

2. Express in simplest form: $\dfrac{1-\dfrac{1}{x}}{1-\dfrac{1}{x^2}}$.

Solution: The LCD of $\frac{1}{x}$ and $\frac{1}{x^2}$ is x^2. Therefore, multiply the numerator *and* the denominator of the complex fraction by 1 in the form of x^2 divided by itself:

$$\left[\frac{x^2}{x^2}\right] \cdot \frac{1-\dfrac{1}{x}}{1-\dfrac{1}{x^2}} = \frac{x^2\left(1-\dfrac{1}{x}\right)}{x^2\left(1-\dfrac{1}{x^2}\right)}$$

Use the distributive property: $\quad = \dfrac{x^2 - \dfrac{x^2}{x}}{x^2 - \dfrac{x^2}{x^2}}$

Simplify: $\quad = \dfrac{x^2 - x}{x^2 - 1}$

Factor and simplify: $\quad = \dfrac{x(x-1)}{(x+1)(x-1)}$

$$= \frac{x}{x+1}$$

Exercise Set 1.5

1–11. Simplify each fraction.

1. $\dfrac{\dfrac{x}{x+1}}{1-\dfrac{x}{x+1}}$

2. $\dfrac{a-1}{\dfrac{1}{a}-1}$

3. $\dfrac{3-\dfrac{3}{x}}{x-1}$

4. $\dfrac{n-\dfrac{1}{n}}{\dfrac{1-n^2}{n}}$

5. $\dfrac{4-x^{-2}}{2-x^{-1}}$

6. $\dfrac{x^{-1}}{x^{-1}+y^{-1}}$

7. $\dfrac{\dfrac{y}{x}-\dfrac{x}{y}}{\dfrac{1}{x}+\dfrac{1}{y}}$

8. $\dfrac{1+\dfrac{b}{a-b}}{1-\dfrac{a}{a-b}}$

9. $\dfrac{\dfrac{a}{b}+1}{\dfrac{a}{b}-\dfrac{b}{a}}$

10. $\dfrac{m - \dfrac{1}{m}}{m - 2 + \dfrac{1}{m}}$ **11.** $\dfrac{a + b^{-1}}{b + a^{-1}}$

12. Express in simplest form the value of $\dfrac{r+s}{1-rs}$ when $r = \dfrac{1}{3}$ and $s = \dfrac{1}{7}$.

1.6 WORKING WITH RADICALS

Square root radicals may be multiplied and divided:

$$2\sqrt{3} \cdot 4\sqrt{7} = 2 \cdot 4\sqrt{3 \cdot 7} = \mathbf{8}\sqrt{\mathbf{21}}.$$

Square root radicals with the same radicand may be combined by adding their numerical coefficients:

$$2\sqrt{5} + 6\sqrt{5} = \mathbf{8}\sqrt{\mathbf{5}}.$$

A square root radical is not considered to be in simplest form until a perfect square factor greater than 1 or a fraction, if any, is removed from its radicand. A fraction that has a radical in its denominator is not considered to be in simplest form until the denominator is made rational.

Using Factoring to Simplify Radicals

If the radicand of a square root radical can be factored so that one of its factors is a perfect square other than 1, then the radical can be simplified by using the fact that

$$\sqrt{ab} = \sqrt{a} \cdot \sqrt{b} \qquad (a, b \ge 0).$$

For example, to simplify $\sqrt{75}$, write the radicand as the product of two numbers such that one of the numbers is the greatest perfect square factor of 75:

$$\sqrt{75} = \sqrt{25 \cdot 3} = \sqrt{25} \cdot \sqrt{3} = \mathbf{5}\sqrt{\mathbf{3}}.$$

Example

1. Write each radical in simplest form:

(a) $\sqrt{72}$ **(b)** $\sqrt{x^5}$ **(c)** $\sqrt{18r^4 s^3}$

Solutions: **(a)** $\sqrt{72} = \sqrt{36}\sqrt{2} = 6\sqrt{2}$

(b) $\sqrt{x^5} = \sqrt{x^4}\sqrt{x} = x^2\sqrt{x}$

(c) $\sqrt{18r^4 s^3} = \sqrt{9r^4 s^2}\sqrt{2s} = 3r^2 s\sqrt{2s}$

Multiplying and Dividing Radicals

Since $\sqrt{ab} = \sqrt{a}\sqrt{b}$, it must also be true that

$$\sqrt{a}\sqrt{b} = \sqrt{ab} \qquad (a,b \geq 0).$$

Thus, to multiply two radicals having the *same index*, write the product of their radicands underneath the same radical sign and then simplify, if possible. Also, note that the product of two identical square root radicals is the radicand:

$$\sqrt{a}\sqrt{a} = \sqrt{a^2} = a .$$

Examples: 1. $\sqrt{8} \cdot \sqrt{6} = \sqrt{48}$

$$= \sqrt{16} \cdot \sqrt{3} = 4\sqrt{3}$$

2. $\left(2\sqrt{3}\right)^2 = \left(2\sqrt{3}\right)\left(2\sqrt{3}\right)$

$$= (2 \cdot 2)\left(\sqrt{3} \cdot \sqrt{3}\right)$$

$$= 4 \cdot 3 = 12$$

Radicals are divided in a similar way since it is true that

$$\frac{\sqrt{a}}{\sqrt{b}} = \sqrt{\frac{a}{b}}, \qquad \text{where } a \geq 0, b > 0$$

Examples: 1. $\dfrac{\sqrt{40}}{\sqrt{8}} = \sqrt{\dfrac{40}{8}} = \sqrt{5}$

2. $\sqrt{\dfrac{3}{4}} = \dfrac{\sqrt{3}}{\sqrt{4}} = \dfrac{\sqrt{3}}{2}$

Combining Radicals

Some radicals must be simplified before they can be combined.

Examples: 1. $5\sqrt{2} + \sqrt{18} = 5\sqrt{2} + \sqrt{9}\sqrt{2}$

$$= 5\sqrt{2} + 3\sqrt{2}$$

$$= (5+3)\sqrt{2} = 8\sqrt{2}$$

2. $2\sqrt{80} - 3\sqrt{20} = 2\sqrt{16}\sqrt{5} - 3\sqrt{4}\sqrt{5}$

$$= 2 \cdot 4\sqrt{5} - 3 \cdot 2\sqrt{5}$$
$$= 8\sqrt{5} - 6\sqrt{5}$$
$$= (8 - 6)\sqrt{5} = \mathbf{2\sqrt{5}}$$

Multiplying Radicals Using FOIL

To find the product of two radical expressions that take the form of a binomial, use FOIL.

Examples:

1. $\left(4 - 2\sqrt{3}\right)^2 = \left(4 - 2\sqrt{3}\right)\left(4 - 2\sqrt{3}\right)$

$$\begin{array}{cccc} F & O & I & L \\ \vee & \vee & \vee & \vee \end{array}$$
$$= 4 \cdot 4 + (4)\left(-2\sqrt{3}\right) + \left(-2\sqrt{3}\right)(4) + \left(-2\sqrt{3}\right)\left(-2\sqrt{3}\right)$$
$$= 16 - 8\sqrt{3} \qquad - 8\sqrt{3} \qquad + (-2)^2\left(\sqrt{3}\right)^2$$
$$= 16 - 16\sqrt{3} \qquad\qquad + 4 \cdot 3$$
$$= \mathbf{28 - 16\sqrt{3}}$$

$$\begin{array}{cccc} F & O & I & L \\ \vee & \vee & \vee & \vee \end{array}$$
2. $\left(7 + 2\sqrt{3}\right)\left(7 - 2\sqrt{3}\right) = 7 \cdot 7 + (7)\left(-2\sqrt{3}\right) + (7)\left(2\sqrt{3}\right) - \left(2\sqrt{3}\right)^2$

$$= 49 - 14\sqrt{3} \qquad + 14\sqrt{3} \qquad - \left(2\sqrt{3}\right)\left(2\sqrt{3}\right)$$
$$= 49 + 0 \qquad\qquad\qquad - 4 \cdot 3$$
$$= \mathbf{37}$$

In Example 2, *conjugate* radical expressions are being multiplied, so their product does not contain a radical. **Conjugate radical expressions** are two radical expressions that take the sum and the difference of the *same* two terms. If A, B, and C are rational and C is greater than 0, then $\left(A + B\sqrt{C}\right)$ and $\left(A - B\sqrt{C}\right)$ are conjugate radical expressions whose product is $A^2 - B^2C$. Thus, the product of two conjugate radical expressions is always rational.

MATH FACTS

PRODUCT OF CONJUGATE RADICAL EXPRESSIONS

$$\left(A + B\sqrt{C}\right)\left(A - B\sqrt{C}\right) = A^2 - B^2$$

Example

2. Multiply: $\left(5 + 3\sqrt{2}\right)\left(5 - 3\sqrt{2}\right)$.

Solution: The factors have the form $A + B\sqrt{C}$ and $A - B\sqrt{C}$, where $A = 5$, $B = 3$, and $C = 2$. Since the product of $A + B\sqrt{C}$ and $A - B\sqrt{C}$ is $A^2 - B^2C$,

$$\left(5 + 3\sqrt{2}\right)\left(5 - 3\sqrt{2}\right) = 5^2 - (3)^2(2)$$
$$= 25 - 9 \cdot 2$$
$$= 7$$

Note: If you forget that $\left(A + B\sqrt{C}\right)\left(A - B\sqrt{C}\right) = A^2 - B^2C$, then multiply using the FOIL method.

Rationalizing Denominators

A fraction whose denominator is a monomial or binomial radical expression can be changed into an equivalent fraction that does not have radicals in its denominator. To do this, multiply the original fraction by a form of 1 that will eliminate the radical denominator.

- For example, to express the fraction $\frac{5}{\sqrt{3}}$ as an equivalent fraction with a rational denominator, multiply the numerator and the denominator of the fraction by $\sqrt{3}$:

$$\frac{5}{\sqrt{3}} = \frac{5}{\sqrt{3}} \cdot \left(\frac{\sqrt{3}}{\sqrt{3}}\right) = \frac{5\sqrt{3}}{\left(\sqrt{3}\right)^2} = \frac{5\sqrt{3}}{3}.$$

- For example, to change the fraction $\frac{7}{7 - 3\sqrt{5}}$ into an equivalent fraction with a rational denominator, multiply the numerator and the denominator of the fraction by the conjugate of $7 - 3\sqrt{5}$, which is $7 + 3\sqrt{5}$:

$$\frac{3}{7 - 3\sqrt{5}} = \frac{3}{7 - 3\sqrt{5}} \cdot \left(\frac{7 + 3\sqrt{5}}{7 + 3\sqrt{5}}\right)$$
$$= \frac{3\left(7 + 3\sqrt{5}\right)}{7^2 - 3^2 5}$$
$$= \frac{3\left(7 + 3\sqrt{5}\right)}{49 - 45}$$
$$= \frac{3\left(7 + 3\sqrt{5}\right)}{4} \text{ or } \frac{21 + 9\sqrt{5}}{4}$$

Examples

3. Combine, and express the result in simplest form: $\dfrac{24}{\sqrt{8}}+\sqrt{32}$

Solution: Rationalize the denominator of the fraction:

$$\frac{24}{\sqrt{8}}+\sqrt{32}=\frac{24}{\sqrt{8}}\cdot\left(\frac{\sqrt{8}}{\sqrt{8}}\right)+\sqrt{32}$$

$$=\frac{24\sqrt{8}}{\left(\sqrt{8}\right)^{2}}\quad+\sqrt{32}$$

Simplify:

$$=\frac{\overset{3}{\cancel{24}}\sqrt{4}\sqrt{2}}{\cancel{8}}+\sqrt{16}\sqrt{2}$$

$$=3\cdot2\sqrt{2}\quad+4\sqrt{2}$$

$$=\mathbf{10\sqrt{2}}$$

4. Express as an equivalent fraction with a rational denominator: $\dfrac{\sqrt{6}}{\sqrt{6}-2}$.

Solution: Multiply the numerator and the denominator by $\sqrt{6}+2$:

$$\frac{\sqrt{6}}{\sqrt{6}-2}\cdot\left(\frac{\sqrt{6}+2}{\sqrt{6}+2}\right)=\frac{\sqrt{6}\left(\sqrt{6}+2\right)}{\left(\sqrt{6}\right)^{2}-2^{2}}$$

$$=\frac{6+2\sqrt{6}}{6-4}=\frac{\overset{1}{\cancel{2}}\left(3+\sqrt{6}\right)}{\cancel{2}}=\mathbf{3+\sqrt{6}}$$

Exercise Set 1.6

1–6. Write each radical in simplest form.

1. $\sqrt{98}$ **3.** $\sqrt{y^{8}}$ **5.** $\sqrt{28r^{5}s}$

2. $2\sqrt{84}$ **4.** $\sqrt{x^{4}y^{3}}$ **6.** $\sqrt[3]{w^{6}y^{3}}$

7–12. Express each product in simplest form.

7. $\sqrt{12}\cdot\sqrt{10}$ **9.** $\left(2\sqrt{3}\right)^{2}$ **11.** $\sqrt{t^{7}}\cdot\sqrt{t^{3}}$

8. $3\sqrt{8}\cdot2\sqrt{50}$ **10.** $\sqrt{x^{5}}\cdot\sqrt{x}$ **12.** $\sqrt{r^{3}}\cdot\sqrt{rs^{2}}$

13–18. Express each sum or difference in simplest form.

13. $2\sqrt{3} + \sqrt{27}$

16. $\sqrt{49x} + \sqrt{x}$

14. $\sqrt{56} - \sqrt{28}$

17. $\sqrt{108y} - \sqrt{12y}$

15. $\sqrt{48} - 2\sqrt{75}$

18. $2\sqrt{20} - 3\sqrt{5} + \sqrt{45}$

19–24. Express each product in simplest form.

19. $\sqrt{3}\left(\sqrt{12} - 2\sqrt{3}\right)$

22. $\left(3 + 2\sqrt{3}\right)\left(3 - \sqrt{12}\right)$

20. $\left(8 - \sqrt{5}\right)\left(8 + \sqrt{5}\right)$

23. $\left(1 - 2\sqrt{3}\right)^2$

21. $\left(\sqrt{3} - \sqrt{7}\right)\left(\sqrt{3} + \sqrt{7}\right)$

24. $\left(\sqrt{2} - \sqrt{3}\right)^2$

25–27. Combine, and express each result in simplest form.

25. $\dfrac{6}{\sqrt{12}} + \sqrt{27}$

26. $\dfrac{15}{\sqrt{20}} - \sqrt{45}$

27. $\dfrac{40}{\sqrt{8}} - \sqrt{50}$

28–33. Express each fraction as an equivalent fraction with a rational denominator.

28. $\dfrac{2}{\sqrt{3} - 1}$

30. $\dfrac{\sqrt{2}}{1 - 3\sqrt{2}}$

32. $\dfrac{1 - \sqrt{3}}{1 + \sqrt{3}}$

29. $\dfrac{24}{1 + \sqrt{7}}$

31. $\dfrac{1}{\sqrt{2} + \sqrt{3}}$

33. $\dfrac{\sqrt{5} + \sqrt{2}}{\sqrt{5} - \sqrt{2}}$

1. Evaluate: $a^0 + a^{-\frac{1}{2}}$, when $a = 9$.

$1 + \dfrac{1}{\sqrt{9}} = 1 + \dfrac{1}{3} = \dfrac{3}{3} + \dfrac{1}{3} = \dfrac{4}{3}$

2. Express as an equivalent fraction with a rational denominator: $\dfrac{2}{5 - 2\sqrt{3}}$.

$\dfrac{2}{5 - 2\sqrt{3}} \cdot \dfrac{(5 + 2\sqrt{3})}{(5 + 2\sqrt{3})} = \dfrac{10 + 4\sqrt{3}}{25 - 4\cdot 3} = \dfrac{10 + 4\sqrt{3}}{13}$

3. Express in simplest form: $\dfrac{a + \dfrac{1}{a}}{a^2 - \dfrac{1}{a}}$.

$\dfrac{a + 1}{a} \cdot \dfrac{a \,(1)}{a^2 - 1} = \dfrac{a + 1}{a^2 - 1} \qquad \dfrac{a + 1}{a(a-1)} \to \dfrac{a + 1}{a - 1}$

(4.) Express the product $(y + 1)\left(\dfrac{y}{1 - y^2}\right)$ as a fraction in lowest terms.

5. Perform the indicated operations, and reduce to lowest terms:

$$\dfrac{x^2 + 2x}{x - 1} \div \dfrac{x^2 - x - 6}{x^2 + 2x - 3}$$

$\dfrac{x(x+3)}{x - 3}$ $\dfrac{x(x+2)}{x-1} \cdot \dfrac{(x+3)(x-1)}{(x+2)(x-3)}$

6. Perform the indicated operations and reduce completely:

$$\dfrac{x^2 + 4x + 3}{x^2 + x} \cdot \dfrac{12 - 3x}{2x} \div \dfrac{x^2 - x - 12}{4x^3}$$

7. Combine and reduce to lowest terms:

$$\dfrac{4x}{2x + 6} + \dfrac{x}{x + 3}.$$

8. The expression $\dfrac{3 + \sqrt{5}}{3 - \sqrt{5}}$ is equivalent to:

(1) $\dfrac{7}{2}$ (2) $\dfrac{7 + 3\sqrt{5}}{7}$ (3) $\dfrac{10\sqrt{5}}{7}$ (4) $\dfrac{7 + 3\sqrt{5}}{2}$

9. Express in simplest form:

$$\dfrac{9}{\sqrt{27}} + \sqrt{12}.$$

27

10. Express as a fraction in simplest form: $\dfrac{\dfrac{1}{n} - \dfrac{1}{3n^2}}{1 - \dfrac{1}{9n^2}}$.

11. Express $\dfrac{5}{4 - \sqrt{13}}$ as an equivalent fraction with a rational denominator.

12. Which fraction is defined for all real numbers?

(1) $\dfrac{x^2 - 1}{(x-1)^2}$　　(2) $\dfrac{x^2 - 1}{x+1}$　　(3) $\dfrac{x^2 - 1}{x^2}$　　(4) $\dfrac{x^2 - 1}{x^2 + 1}$

ANSWERS TO SELECTED EXERCISES: CHAPTER 1

Section 1.1

1. $\dfrac{3}{y^2}$

3. $\dfrac{y^2}{3}$

5. $\dfrac{1}{x^4 y^6}$

7. (a) $1000m$

　　(b) m^3

　　(c) $\dfrac{m}{100}$

　　(d) $\sqrt{m}$

9. $\dfrac{q^6}{p^9}$

11. $\dfrac{2a^8}{3b^3}$

13. $\frac{13}{8}$

15. 8

17. 16

19. $\frac{9}{16}$

21. $\frac{1}{512}$

23. $\frac{64}{125}$

25. $\frac{17}{16}$

27. y^5

29. $3x^4$

31. $-2x^2 y^{\frac{2}{3}}$

Section 1.2

1. $6x^2 y(4x + 5y^4)$

3. Not factorable

5. $x(x^2 - x + 1)$

7. $8u^2 w^2(u^3 - 5w^3)$

9. $x(x + a)$

11. $\left(\dfrac{x}{2} - 6\right)\left(\dfrac{x}{2} + 6\right)$

13. $(t + 7)(t - 3)$

15. $(2q + 5)(q - 3)$

17. $(5s - 1)(s + 3)$

19. $2(y - 5)(y + 5)$

21. $-(x + 9)(x - 5)$

23. $-5(t - 1)(t + 1)$

25. $p(2p - 5)(p + 3)$

27. $(c - 9)(c + 5)$

29. $x = 5$

31. $\dfrac{3}{x}$

33. $\dfrac{-1}{a + b}$

35. $\dfrac{x + y}{x - y}$　　**37.** 0 and 2

Section 1.3

1. $\dfrac{6(x-7)}{xy^3}$

3. $\dfrac{2(x-4)}{(x-5)(x-2)}$

5. $\dfrac{5}{x+7}$

7. $\dfrac{x+3}{x(x-3)}$

9. $\dfrac{3p}{(p-3)(p+2)}$

11. $\dfrac{-r(r+s)}{s}$

13. $\dfrac{-4}{9(t+2)}$

Section 1.4

1. 3

3. $a+b$

5. $\dfrac{b+12}{b(b-3)}$

7. $\dfrac{x+13}{2(x-1)(x+1)}$

9. $\dfrac{a^2+2ab-b^2}{(a-b)(a+b)}$

11. $\dfrac{5(x+1)}{(x+3)(x-2)(x+2)}$

13. $\dfrac{2(r+4s)}{(r-s)(r+s)^2}$

15. $\dfrac{x^2-2x-6}{(x-3)(x+3)(x-2)}$

17. -2

Section 1.5

1. x

3. $\dfrac{3}{x}$

5. $\dfrac{2x+1}{x}$

7. $y-x$

9. $\dfrac{a}{a-b}$

11. $\dfrac{a}{b}$

Section 1.6

1. $7\sqrt{2}$

3. y^4

5. $2r^2\sqrt{7s}$

7. $2\sqrt{30}$

9. 12

11. t^5

13. $5\sqrt{3}$

15. $-6\sqrt{3}$

17. $4\sqrt{3y}$

19. 0

21. -4

23. $13-4\sqrt{3}$

25. $4\sqrt{3}$

27. $5\sqrt{2}$

29. $-4\left(1-\sqrt{7}\right)$

31. $\sqrt{3}-\sqrt{2}$

33. $\dfrac{7+2\sqrt{10}}{3}$

35. $-\dfrac{4+\sqrt{6}}{10}$

Regents Tune-Up: Chapter 1

1. $\frac{4}{3}$

2. $\dfrac{2\left(5+2\sqrt{3}\right)}{13}$

3. $\dfrac{1}{a-1}$

4. $\dfrac{y}{1-y}$

5. $\dfrac{x(x+3)}{x-3}$

6. $-6x$

7. $\dfrac{3x}{x+3}$

8. (4)

9. $3\sqrt{3}$

10. $\dfrac{3}{3n+1}$

11. $\dfrac{5\left(4+\sqrt{13}\right)}{3}$

12. (4)

CHAPTER 2

REVIEW AND EXTENSION OF EQUATION SOLVING

2.1 SOLVING ABSOLUTE-VALUE EQUATIONS AND INEQUALITIES

KEY IDEAS

The **absolute value** of a real number x, written as $|x|$, represents the *distance* of that number from 0 on a number line. Since distance cannot be negative,

$$|x| = x \text{ when } x \geq 0 \quad \text{and} \quad |x| = -x \text{ when } x < 0.$$

Equations and inequalities that involve absolute value are solved by writing equivalent statements using the following rules:

- If $|x| = d$, then $x = -d$ or $x = d$.

  Graph:
- If $|x| < d$, then $-d < x < d$.

  Graph:
- If $|x| > d$, then $x < -d$ or $x > d$.

  Graph:

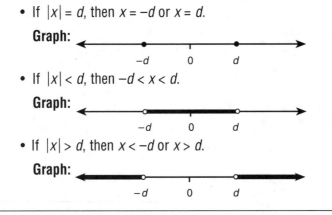

Solving Absolute-Value Equations

To solve a first-degree absolute-value equation of the form $|ax - b| = d$, write and then solve each equation in the disjunction:

$$ax - b = -d \quad \text{or} \quad ax - b = d.$$

Examples

1. Find the solution set: $|2x + 3| = 1$.

Solution: If $|2x+3|=1$, then $2x+3=-1$ or $2x+3=1$

$$2x=-4 \qquad\qquad 2x=-2$$

$$\frac{2x}{2}=\frac{-4}{2} \qquad\qquad \frac{2x}{2}=\frac{-2}{2}$$

$$x=-2 \quad \text{or} \quad x=-1$$

The solution set is $\{-2,-1\}$.

2. Solve and check: $|x+3|=2x$.

Solution: If $|x+3|=2x$, then $x+3=-2x$ or $x+3=2x$

$$3x+3=0 \qquad\qquad 3=x$$

$$3x=-3 \qquad\qquad x=3$$

$$x=-1$$

To check x = −1:   *To check* x = 3:

$$|x+3|=2x \qquad\qquad |x+3|=2x$$

$$|-1+3| \; ? \; 2(-1) \qquad\qquad |3+3| \; ? \; 2\cdot 3$$

$$|2| \; ? \; {-2} \qquad\qquad |6| \; ? \; 6$$

$$2 \neq -2 \qquad\qquad 6=6$$

Hence the only root of $|x+3|=2x$ is $x=3$.

Solving Absolute-Value Inequalities (Less Than)

To solve a first-degree absolute-value inequality of the form $|ax-b|<d$, write the compound inequality

$$-d<ax-b<d$$

and then solve for x. If $|ax-b|\le d$, remove the absolute-value sign by writing

$$-d\le ax-b\le d.$$

Example

3. Solve for x and graph the solution set:

(a) $|2x-1|\le 7$ **(b)** $|2x-1|<7$

Solutions: **(a)** If $|2x-1|\le 7$, then

$$-7\le 2x-1 \qquad \le 7$$

Add 1 to each member: $+1-7\le 2x-1+1\le 7+1$

Simplify: $-6\le 2x \qquad\qquad \le 8$

Divide each member by 2: $\dfrac{-6}{2}\le\dfrac{2x}{2} \qquad\qquad \le\dfrac{8}{2}$

$$-3\le x \qquad\qquad \le 4$$

The solution interval is $-3 \le x \le 4$. The graph is shown below.

As a check you should pick any convenient point in this interval, for example, 0, and verify that it satisfies the original absolute-value inequality.

(b) If $|2x - 1| < 7$, then $-7 \le 2x - 1 < 7$. Using the result obtained in part (a), you find that the solution interval is $-3 < x < 4$. The graph is shown below.

Solving Absolute-Value Inequalities (Greater Than)

To solve a first-degree absolute-value inequality of the form $|ax - b| > d$, write and then solve the equations in the disjunction

$$ax - b < -d \quad \text{or} \quad ax - b > d.$$

If $|ax - b| \ge d$, then remove the absolute-value sign by writing the disjunction

$$ax - b \le -d \quad \text{or} \quad ax - b \ge d.$$

Example

4. Solve for x and graph the solution set:
(a) $|3x + 1| \ge 7$ (b) $|3x + 1| > 7$

Solutions: **(a)** If $|3x + 1| \ge 7$, then

$$3x + 1 \le -7 \quad \text{or} \quad 3x + 1 \ge 7.$$

Solve each inequality:

$$
\begin{array}{c|c}
3x \le -8 & 3x \ge 6 \\
\dfrac{3x}{3} \le -\dfrac{8}{3} & \dfrac{3x}{3} \ge \dfrac{6}{3} \\
x \le -\dfrac{8}{3} & x \ge 2
\end{array}
$$

The solution intervals are $x \le \frac{8}{3}$ and $x \ge 2$. The graph is shown below.

(b) If $|3x + 1| > 7$, then $3x + 1 < -7$ or $3x + 1 > 7$. Using the result obtained in part (a), you find that the solution intervals are $x < -\frac{8}{3}$ and $x > 2$. The graph is shown below.

Exercise Set 2.1

1–15. Find and then graph the solution set.

1. $|x| = 5$
2. $|x| + 4 = 0$
3. $|x - 3| \le 2$
4. $4 > |x + 1|$
5. $\dfrac{|x + 1|}{2} > 3$

6. $|2 - y| \le 1$
7. $|2x - 5| \le 7$
8. $|3 - x| > 1$
9. $|4x + 1| \ge 9$
10. $\dfrac{|2x - 1|}{3} > 5$

11. $3 + |x + 5| \le 6$
12. $-2|4x - 1| > 10$
13. $|3 - 2x| < 7$
14. $11 < |4 - 3x|$
15. $\dfrac{|2 - n|}{5} - 1 \le 0$

16 and 17. Solve for y *and check.*

16. $|1 - y| = y + 2$ 17. $|y + 3| \le 2y$

2.2   SOLVING QUADRATIC INEQUALITIES ALGEBRAICALLY

KEY IDEAS

Inequalities such as

$$x^2 - 2x - 3 < 0 \quad \text{and} \quad x^2 - 2x - 3 > 0$$

are **quadratic inequalities** that are in standard form since the nonzero terms appear on the left side of the inequality symbol. To solve either inequality, factor the quadratic polynomial.

The set of all values of x that make the two factors have different signs, so that their product is negative, is the solution to $x^2 - 2x - 3 < 0$. The set of all values of x that make the two binomial factors have the same sign, so that their product is positive, is the solution to $x^2 - 2x - 3 > 0$.

If the inequality is "$\ge$" or "$\le$," the solution set also includes the values of x for which the quadratic polynomial is 0.

Solving Quadratic Inequalities by Factoring

To solve a quadratic inequality such as $x^2 - 2x < 3$:

- Write the quadratic inequality in standard form and factor:

$$x^2 - 2x - 3 < 0$$
$$(x + 1)(x - 3) < 0$$

33

- Find the roots of the related quadratic equation. If $(x+1)(x-3)=0$, then

$$x = -1 \quad \text{or} \quad x = 3.$$

- Locate the roots on a number line. The two roots divide the number line (Figure 2.1) into three possible solution intervals:
  (I) $x < -1$,
  (II) $-1 < x < 3$,
  and (III) $x > 3$.

Figure 2.1 Locating Roots on a Number Line

- Determine the sign of the product of the two factors on each interval.

Interval	Select a Test Value	Values of Factors		$(x+1)(x-3)<0$?
		$(x+1)$	$(x-3)$	
$x<-1$	$x=-2$	-1	-5	No, since $(-1)(-5)>0$.
$-1<x<3$	$x=0$	1	-3	Yes, since $(1)(-3)<0$.
$x>3$	$x=4$	5	1	No, since $(5)(1)>0$.

- Write and graph the solution set. The solution set is $\{x: -1 < x < 3\}$. The graph is shown below.

General Form of the Solution Intervals

If r_1 and r_2 $(r_1 < r_2)$ represent the roots of $ax^2+bx+c=0\,(a>0)$, then the solution intervals to the related set of quadratic inequalities are given in Table 2.1. Knowing these general forms can greatly simplify matters. For example, the solution interval to the quadratic inequality $x^2+x-6<0$ has the form $r_1 < x < r_2$, where r_1 and r_2 are the roots of the related quadratic equation $x^2+x-6=0$. You can verify that the roots of this equation are $x=-3$ and $x=2$. Since -3 is less than 2, let $r_1 = -3$ and $r_2 = 2$. The solution interval is, therefore, $-3 < x < 2$. If the original quadratic inequality was $x^2+x-6\geq0$, the solution intervals would be $x \leq -3$ or $x \geq 2$.

TABLE 2.1 GENERAL SOLUTIONS TO QUADRATIC INEQUALITIES ($a > 0$)

Inequality	Solution Interval	Graph
$ax^2 + bx + c < 0$	$r_1 < x < r_2$	![graph] r_1 r_2
$ax^2 + bx + c \leq 0$	$r_1 \leq x \leq r_2$	![graph] r_1 r_2
$ax^2 + bx + c > 0$	$x < r_1$ or $x > r_2$	![graph] r_1 r_2
$ax^2 + bx + c \geq 0$	$x \leq r_1$ or $x \geq r_2$	![graph] r_1 r_2

Examples

1. Solve, and graph the solution set: $8 - 2x \leq x^2$.

Solution: $\qquad\qquad\qquad\qquad 8 - 2x \leq x^2$

Collect terms on the same side: $\qquad 0 \leq x^2 + 2x - 8$

Write in standard form: $\qquad x^2 + 2x - 8 \geq 0$

Write the related equation: $\qquad x^2 + 2x - 8 = 0$

Solve by factoring: $\qquad (x + 4)(x - 2) = 0$

$\qquad\qquad\qquad\qquad\qquad x = -4 \quad$ or $\quad x = 2$

Let $r_1 = -4$ and $r_2 = 2$. The general form of the solution to $x^2 + 2x - 8 \geq 0$ is $x \leq r_1$ or $x \geq r_2$. Hence the solution set is $\{x : x \leq -4$ or $x \geq 2\}$. The graph is shown below.

$\qquad\qquad\qquad\qquad -4 \qquad\qquad 2$

2. Solve, and graph the solution set: $4x > x^2$.

Solution: If $4x > x^2$, then $0 > x^2 - 4x$ or, equivalently, $x^2 - 4x < 0$. Find r_1 and r_2, the roots of the related quadratic equation.

$$x^2 - 4x = 0$$
$$x(x - 4) = 0$$
$$x = 0 \quad \text{or} \quad x - 4 = 0$$
$$x = 0 \quad | \quad\quad x = 4$$

The solution interval for $x^2 - 4x < 0$ has the form $r_1 < x < r_2$. Since $r_1 = 0$ and $r_2 = 4$, the solution set is $\{x : 0 < x < 4\}$. The graph is shown below.

$\qquad\qquad\qquad\qquad 0 \qquad\qquad 4$

Exercise Set 2.2

1–15. Solve, and graph the solution set.

1. $x^2 + 4x - 5 \geq 0$ **6.** $x^2 + 4 \leq 4x$ **11.** $4m \leq m^2$

2. $n^2 - 6n \leq 27$ **7.** $7x \geq x^2$ **12.** $6h^2 < 7h + 3$

3. $x^2 - 5x \geq 0$ **8.** $2x^2 > 5x + 3$ **13.** $0 \leq -(x^2 + 2x - 63)$

4. $y^2 + 9y \leq 36$ **9.** $6 \leq t^2 - t$ **14.** $(x - 2)(x + 1) \geq 10$

5. $b^2 - 25 \leq 0$ **10.** $x^2 - 4x + 4 \geq 0$ **15.** $x^2 - 6x + 9 < 0$

16. When a baseball is hit by a batter, the height of the ball, h, at time t ($t \geq 0$) is determined by the equation $h = -16t^2 + 64t + 4$. For which interval of time is the height of the ball greater than or equal to 52 feet?

2.3 SOLVING FRACTIONAL EQUATIONS

⌃ KEY IDEAS

An equation such as

$$\frac{2}{y} - \frac{9}{10} = \frac{1}{5y}$$

is called a **fractional equation** since at least one of the fractions in the equation contains a variable in the denominator. A fractional equation can be cleared of its fractions by multiplying each term of the equation, right and left sides, by the lowest common denominator (LCD) of its fractions.

The solution of the new equation may include *extra* roots that are *not* roots of the fractional equation. These extra roots are called **extraneous roots**. Extraneous roots may be introduced whenever the original equation is multiplied by a variable expression.

To solve the fractional equation

$$\frac{2}{y} - \frac{9}{10} = \frac{1}{5y}$$

multiply each term by the LCD, $10y$, and then solve the resulting equation.

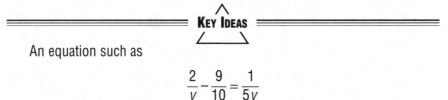

$$20 - 9y = 2$$
$$-9y = 2 - 20$$
$$y = \frac{-18}{-9}$$
$$= 2$$

It is left to you to check that $y = 2$ is a root.

Example

Solve for x and check:

$$\frac{x}{x+2} - \frac{1}{x-2} = \frac{8}{x^2 - 4}.$$

Solution: Rewrite the equation with its denominator(s) factored:

$$\frac{x}{x+2} - \frac{1}{x-2} = \frac{8}{(x+2)(x-2)}$$

The LCD of the denominators is $(x-2)(x+2)$. Hence each term of the fractional equation must be multiplied by $(x-2)(x+2)$:

$$(x-2)(x+2)\left(\frac{x}{x+2}\right) - (x-2)(x+2)\left(\frac{1}{x-2}\right) = (x-2)(x+2)\left[\frac{8}{(x+2)(x-2)}\right]$$

$$(x-2)\cancel{(x+2)}\frac{\overset{1}{x}}{\cancel{x+2}} - \cancel{(x-2)}(x+2)\frac{\overset{1}{1}}{\cancel{x-2}} = \cancel{(x-2)}\cancel{(x+2)}\frac{8}{\cancel{(x+2)}\cancel{(x-2)}}$$

$$x(x-2) - (x+2) = 8$$
$$x^2 - 2x - x - 2 = 8$$
$$x^2 - 3x - 10 = 0$$
$$(x+2)(x-5) = 0$$
$$x + 2 = 0 \quad \text{or} \quad x - 5 = 0$$
$$x = -2 \quad | \quad x = 5$$

Check: If $x = -2$, then denominators $x + 2$ and $x^2 - 4$ in the original equations will be equal to 0, which is not permitted. Hence $x = -2$ is an *extraneous root* and is, therefore, rejected. You should verify that $x = 5$ satisfies the original fractional equation. The fractional equation has the single root $x = 5$.

Exercise Set 2.3

1–16. Solve for the variable and check.

1. $\dfrac{1}{5} - \dfrac{1}{3x} = \dfrac{1}{15x}$

2. $\dfrac{3}{5r} - \dfrac{2}{6r} = \dfrac{1}{15}$

3. $\dfrac{1}{x} + \dfrac{3}{2x} = \dfrac{x-4}{2}$

4. $\dfrac{5}{w} + 3w = \dfrac{17}{w}$

5. $\dfrac{x+2}{4x} - \dfrac{1}{x} = \dfrac{1}{12}$

6. $\dfrac{1}{y} - \dfrac{y+1}{8} = \dfrac{y-1}{4y}$

7. $\dfrac{x+15}{5x} - \dfrac{1}{2} = \dfrac{x-2}{2x}$

8. $\dfrac{3}{5} + \dfrac{n-2}{3} = \dfrac{14}{5n}$

9. $\dfrac{x}{5} - \dfrac{6}{x+2} = \dfrac{7x-16}{5x+10}$

10. $\dfrac{t}{t-3} + \dfrac{t-2}{2} = \dfrac{5t-3}{4t-12}$

11. $\dfrac{8}{y^2-25} + \dfrac{2}{y+5} = \dfrac{y}{y-5}$

12. $\dfrac{2}{n-3} + \dfrac{n-2}{n^2-2n-3} = \dfrac{2n+1}{n+1}$

13. $\dfrac{1}{b-3} - \dfrac{3}{2b+6} = \dfrac{b}{b^2-9}$

14. $\dfrac{1}{x-2} + \dfrac{x+2}{x+5} = \dfrac{3}{x^2+3x-10}$

15. $\dfrac{4}{m^2-16} + \dfrac{m-3}{m+4} = \dfrac{1}{m-4}$

16. $\dfrac{2}{r-6} = \dfrac{10}{r^2-7r+6} - \dfrac{r}{r-1}$

2.4 SOLVING RADICAL EQUATIONS AND EQUATIONS WITH RATIONAL EXPONENTS

KEY IDEAS

Equations such as
$$\sqrt{2x+3} = 7, \quad \sqrt{1-3x} = 8, \quad \text{and} \quad \sqrt{x+2} - 3 = 2x$$
are examples of **radical equations**, since the variable is included in the radicand. An equation having a square root radical is solved by isolating the radical and then squaring both sides of the equation. If the radical is a cube root, it is eliminated by cubing both sides of the equation.

If $x^{\frac{p}{k}} = c$, where c is some nonzero constant, then
$$x = c^{\frac{k}{p}}.$$

Solving Radical Equations

An equation containing a radical is solved by isolating the radical and then raising *both* sides of the equation to the power that eliminates the radical. Since this procedure may introduce *extraneous roots*, be sure to check all roots of the transformed equation in the original radical equation.

Examples

1. Solve for n: $\qquad \sqrt{2n} = 4\sqrt{5}$.

Solution: Square both sides:
$$\left(\sqrt{2n}\right)^2 = \left(4\sqrt{5}\right)^2$$
$$2n = 4^2\left(\sqrt{5}\right)^2$$

38

$$2n = 16 \cdot 5$$
$$n = \frac{80}{2}$$
$$= \mathbf{40} \qquad \text{The check is left for you.}$$

2. Solve for x: $\sqrt{4-x} - 8 = x$.

Solution:

$$\sqrt{4-x} - 8 = x$$

Isolate the radical: $\qquad\qquad\qquad \sqrt{4-x} = x + 8$

Square both sides: $\qquad\qquad\quad \left(\sqrt{4-x}\right)^2 = (x+8)^2$

$$4 - x = x^2 + 16x + 64$$

Combine like terms: $\qquad\qquad 0 = x^2 + 16x + x + 64 - 4$

Write in standard form: $\qquad x^2 + 17x + 60 = 0$

Factor: $\qquad\qquad\qquad\qquad (x+5)(x+12) = 0$

$$x + 5 = 0 \quad \text{or} \quad x + 12 = 0$$
$$x = -5 \quad | \quad\qquad x = -12$$

To check x = −5:

$$\sqrt{4-x} - 8 \overset{?}{=} x$$
$$\sqrt{4-(-5)} - 8 \;\Big|\; -5$$
$$\sqrt{4+5} - 8 \;\Big|\; -5$$
$$3 \quad -8 \;\Big|\; -5$$
$$-5 \quad = -5$$

To check x = −12:

$$\sqrt{4-x} - 8 \overset{?}{=} x$$
$$\sqrt{4-(-12)} - 8 \;\Big|\; -12$$
$$\sqrt{4+12} - 8 \;\Big|\; -12$$
$$4 \quad -8 \;\Big|\; -12$$
$$-4 \quad \neq -12$$

Therefore, −5 is a root and −12 is an *extraneous root* that must be rejected. The solution of $\sqrt{4-x} - 8 = x$ is the single root $x = \mathbf{-5}$.

3. Solve for x: $\sqrt[3]{3x-4} = 2$.

Solution: Raise both members of the equation to the third power:

$$\left(\sqrt[3]{3x-4}\right)^3 = 2^3$$
$$3x - 4 = 8$$
$$3x = 12$$
$$x = \mathbf{4} \qquad \text{The check is left for you.}$$

Solving Equations with Rational Exponents

Equations in which the variable has a rational exponent of the form $\frac{p}{k}$ are solved by isolating the variable and then raising both sides of the equation to the $\frac{k}{p}$ power.

Example

4. Solve for x:

(a) $x^{\frac{3}{2}} - 3 = 5$

(b) $x^{-\frac{4}{3}} = \dfrac{81}{16}$

Solutions: **(a)** $x^{\frac{3}{2}} - 3 = 5$

$$x^{\frac{3}{2}} = 8$$

$$\left(x^{\frac{3}{2}}\right)^{\frac{2}{3}} = 8^{\frac{2}{3}}$$

$$x = 4$$

(b) $x^{-\frac{4}{3}} = \dfrac{81}{16}$

$$\left(x^{-\frac{4}{3}}\right)^{-\frac{3}{4}} = \left(\dfrac{81}{16}\right)^{-\frac{3}{4}}$$

$$x = \left(\dfrac{16}{81}\right)^{\frac{3}{4}}$$

$$= \left(\dfrac{2}{3}\right)^3 = \dfrac{8}{27}$$

Exercise Set 2.4

1–15. Solve for the variable and check.

1. $\sqrt{x^2 - 25} = x - 1$

2. $\sqrt{5x + 1} = 25 - 3x$

3. $\sqrt{x - 1} = x - 7$

4. $\sqrt{3 - x} = 3 + x$

5. $2\sqrt{3y} = 5\sqrt{12}$

6. $\sqrt{13 - 3x} + x = 1$

7. $n - \sqrt{3n + 4} = 2$

8. $\sqrt{1 - 2x} = \sqrt{x^2 - 7}$

9. $2\sqrt{1 - 3p} = 3 - p$

10. $t - 2\sqrt{4t - 7} = 0$

11. $x^{\frac{2}{3}} = 25$

12. $2x^{\frac{3}{2}} = 54$

13. $x^{-\frac{4}{3}} = 16$

14. $x^{-\frac{3}{4}} = \dfrac{27}{8}$

15. $x^{-\frac{2}{3}} = 16^{-\frac{1}{2}}$

REGENTS TUNE-UP: CHAPTER 2

Each of the questions in this section has appeared on a previous Course III Regents Examination. Here is an opportunity for you to review Chapter 2 and, at the same time, prepare for the Course III Regents Examination.

1. What is the solution set of the equation $\sqrt{9x^2 - 11} = 5$?

2. Solve for x: $x^{\frac{3}{2}} = 64$.

3. What is the solution set of the equation $2x - |x+3| = 9$?
(1) $\{12\}$　　(2) $\{2\}$　　(3) $\{2, 12\}$　　(4) $\{\ \}$

4. What is the solution set of the inequality $x^2 - 6x - 7 > 0$?

5. Solve for x: $\sqrt{x^2 + 7} = x + 1$.

6. What is the solution set of $|3 - 2x| = 5$?

7. What is the solution set of $(x+3)(x-2) > 0$?

(1) 　　(3)

(2) 　　(4)

8. What is the graph of the solution set of $|2x - 1| < 9$?

(1) 　　(3)

(2) 　　(4)

9. What is the solution set of $x^2 - x - 6 < 0$?
(1) $\{x \mid x < -2 \text{ or } x > 3\}$　　(3) $\{x \mid -3 < x < 2\}$
(2) $\{x \mid x < -3 \text{ or } x > 2\}$　　(4) $\{x \mid -2 < x < 3\}$

10. What is the solution set of the equation $|3x - 1| = x + 5$?
(1) $\{-1\}$　　　　　　　(3) $\{3\}$
(2) $\{-1, 3\}$　　　　　　(4) $\{1, -3\}$

11. What is the solution set of $x^2 - 3x - 28 \geq 0$?
(1) $x \geq 7$ or $x \leq -4$　　(3) $-4 \leq x \leq 7$
(2) $x \leq 7$ or $x \geq -4$　　(4) $-4 < x < 7$

41

12. Solve for x:

$$\frac{x}{x-2} - \frac{8}{x+3} = \frac{10}{x^2+x-6}.$$

ANSWERS TO SELECTED EXERCISES: CHAPTER 2

Section 2.1

1. $\{-5, 5\}$

2. $\{\ \}$

3. $\{x: 1 \le x \le 5\}$

4. $\{x: -5 < x < 3\}$

5. $\{x: x < -7 \text{ or } x > 5\}$

6. $\{y: 1 \le y \le 3\}$

7. $\{x: -1 < x \le 6\}$

8. $\{x: x < 2 \text{ or } x > 4\}$

9. $\{x: x \le -\frac{5}{2} \text{ or } x \ge 2\}$

10. $\{x: x < -7 \text{ or } x > 8\}$

11. $\{x: -8 \le x \le -2\}$

12. $\{\ \ \}$

13. $\{x: -2 < x < 5\}$

14. $\{x: x < -\frac{7}{3} \text{ or } x > 5\}$

15. $\{n: -3 \le n \le 7\}$

16. $-\frac{1}{2}$

17. $y \ge 3$

Section 2.2

1. $x \le -5$ or $x \ge 1$

2. $-3 \le n \le 9$

3. $x \le 0$ or $x \ge 5$

4. $-12 \le y \le 3$

5. $-5 \le b \le 5$

6. 2

7. $0 \le x \le 7$

8. $x < -\frac{1}{2}$ or $x > 3$

9. $t \le -2$ or $t \ge 3$

10. All real numbers

11. $m \le 0$ or $m \ge 4$

12. $-\frac{1}{3} < h < \frac{3}{2}$

13. $x < -9$ or $x \ge 7$

14. $x \le -3$ or $x \ge 4$

15. No real numbers

16. $1 \le t \le 3$

Section 2.3

1. 2

3. -1 or 5

5. 3

7. $\frac{5}{2}$

9. 7

11. -2 or -1

13. 5

15. 2 or 6

Section 2.4

1. 13

3. 10

5. 25

7. 7

9. -5 or -1

11. 125

12. 9

13. $\frac{1}{8}$

14. $\frac{16}{81}$

15. 8

Regents Tune-Up: Chapter 2

1. -2 or 2

2. 16

3. (1)

4. $x < -1$ or $x > 7$

5. 3

6. -1 or 4

7. (4)

8. (2)

9. (4)

10. (2)

11. (1)

12. 3

CHAPTER 3

COMPLEX NUMBERS

3.1 CLASSIFYING NUMBERS

$$\bigwedge_{\text{KEY IDEAS}}$$

The numbers of arithmetic—whole numbers, integers, rational numbers (fractions), and irrational numbers—form the set of **real numbers**. A number such as $\sqrt{-4}$ is *not* real. A new type of number, called *imaginary*, was invented to make it possible to work with square roots of negative numbers. The set of **imaginary numbers** is based on the **imaginary unit *i***, which is so defined that

$$i^2 = -1 \text{ and } i = \sqrt{-1}.$$

Square roots of negative numbers other than -1 can be expressed in terms of i by factoring out $\sqrt{-1}$ and replacing it with i. For example,

$$\sqrt{-4} = \sqrt{4} \cdot \sqrt{-1} = 2i.$$

The sum of a real number and an imaginary number, such as $3 + 2i$, is called a **complex number**.

Field Properties

A set of numbers S and two operations, such as addition and multiplication, form a **field** (**F**) if each of the field properties listed in Table 3.1 is true for any numbers a, b, and c that are in set S.

TABLE 3.1 FIELD PROPERTIES

Field Property	Addition	Multiplication
F_1: Closure	$a + b$ is in S.	$a \cdot b$ is in S.
F_2: Commutative	$a + b = b + a$	$a \cdot b = b \cdot a$
F_3: Associative	$(a + b) + c = a + (b + c)$	$(a \cdot b)c = a(b \cdot c)$
F_4: Identity	0 is in S: $a + 0 = 0 + a = a$	1 is in S: $a \cdot 1 = 1 \cdot a = a$
F_5: Inverse	$-a$ is in S: $a + (-a) = (-a) + a = 0$	$\dfrac{1}{a}$ is in S $(a \neq 0)$ $a \cdot \dfrac{1}{a} = \dfrac{1}{a} \cdot a = 1$
F_6: Distributive	$a(b + c) = ab + ac$ and $(b + c)a = ba + ca$	

Properties F_1 through F_6 in Table 3.1 hold for *all* real numbers so that the set of real numbers is a field. A subset of the real numbers, however, may not be a field. For example, to determine whether the set of integers forms a field under the operations of addition and multiplication, determine whether each of the field properties holds:

F_1: The set of integers is closed under addition and multiplication because the sum or product of any two integers is an integer.

F_2: For any two integers, the order in which they are added or multiplied does not matter, so the commutative property holds.

F_3: For any three integers, the way in which the numbers are grouped when added or multiplied does not matter, so the associative property holds.

F_4: Adding the integer 0 to any integer a gives a, so that the additive identity is 0. Multiplying any integer a by the integer 1 gives a, so that the multiplicative identity is 1.

F_5: Adding a and its opposite, $-a$, gives the identity 0, so that each integer has an additive inverse. However, each nonzero integer does *not* have a multiplicative inverse (reciprocal) that is an integer. For example, if $a = 2$, then $2(\frac{1}{2}) = 1$. Since $\frac{1}{2}$ is not an integer, $a = 2$ does *not* have a multiplicative inverse.

Hence the set of integers is *not* a field, since it lacks property F_5 in Table 3.1 for the operation of multiplication.

The Set of Complex Numbers

An even root of a negative real number is called a **pure imaginary number**. The square roots of negative numbers may be expressed in terms of the imaginary unit i, where $i = \sqrt{-1}$.

> *Examples:* 1. $\sqrt{-9} = \sqrt{9}\sqrt{-1} = 3i$
> 2. $-\sqrt{-6} = -\sqrt{6}\sqrt{-1} = -i\sqrt{6}$
> 3. $\sqrt{-50} = \sqrt{25}\sqrt{-2} = 5\sqrt{2}\sqrt{-1} = 5i\sqrt{2}$

A **complex number** is a number that can be written in standard $a + bi$ form, where a and b are real numbers, and $i = \sqrt{-1}$. For example, $\frac{2+i}{3}$, 5, and $4i$ are complex numbers since:

- $\frac{2+i}{3} = \frac{2}{3} + \frac{1}{3}i$. Here, $a = \frac{2}{3}$ and $b = \frac{1}{3}$. A complex number represents the indicated sum of a real number and an imaginary number.

- $5 = 5 + 0i$. Here, $a = 5$ and $b = 0$. Since every real number is also a complex number, the set of real numbers is a *subset* of the set of complex numbers.

- $4i = 0 + 4i$. Here, $a = 0$ and $b = 4$. The set of imaginary numbers is a *subset* of the set of complex numbers.

Arithmetic operations involving complex numbers are defined so that the set of complex numbers is a field. In performing arithmetic operations, terms involving i are treated as if they are monomials. Here are four examples:

(1) $3i + 5i = 8i$;   (2) $6i - i = 5i$;   (3) $i \cdot i = i^2$;   (4) $i^{13} \div i^4 = i^9$.

Examples

1. Find the sum of $2 + 3i$ and $4 - 5i$.

Solution: The sum of two complex numbers is the sum of their real parts and the sum of their imaginary parts. Treat the imaginary parts of complex numbers as monomials involving i:

$$(2 + 3i) + (4 - 5i) = (2 + 4) + (3i - 5i) = \mathbf{6 - 2i}.$$

2. Subtract $-4 + 6i$ from $1 - 2i$.

Solution: $(1 - 2i) - (-4 + 6i) = (1 - 2i) + (4 - 6i)$
$$= (1 + 4) + (-2i - 6i) = \mathbf{5 - 8i}$$

3. Find the additive inverse of $-1 + 2i$.

Solution: The sum of a complex number and its additive inverse is 0. Since $(-1 + 2i) + (1 - 2i) = 0$, the additive inverse of $-1 + 2i$ is $\mathbf{1 - 2i}$.

4. Express as a monomial in terms of i:

(a) $\sqrt{-45}$ **(b)** $3\sqrt{-16} + \sqrt{-36}$ **(c)** $\sqrt{-18} + \sqrt{-32}$

Solutions: Factor out $\sqrt{-1}$ from each radical and simplify.

(a) $\sqrt{-45} = \sqrt{45}\sqrt{-1} = \sqrt{9}\sqrt{5}i = \mathbf{3\sqrt{5}i}$

(b) $3\sqrt{-16} + \sqrt{-36} = 3\sqrt{16}\sqrt{-1} + \sqrt{36}\sqrt{-1}$

$$= 12i \qquad + 6i = \mathbf{18i}$$

(c) $\sqrt{-18} + \sqrt{-32} \quad = \sqrt{9}\sqrt{2}i \quad + \sqrt{6}i\sqrt{2}i$

$$= 3\sqrt{2}i \qquad + 4\sqrt{2}i = \mathbf{7\sqrt{2}i}$$

Graphing Complex Numbers

A complex number of the form $a + bi$ can be graphed in the *complex* coordinate plane by using the ordered pair (a, b). The value of a is measured horizontally along the **real axis**, and the value of b is measured vertically along the **imaginary axis**. In Figure 3.1, the complex number $3+5i$ is graphed by drawing a directed segment from the origin O to the ordered pair (3, 5).

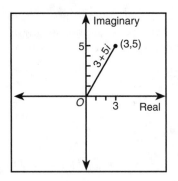

Figure 3.1 Graphing a Complex Number

Example

5. In which quadrant is the sum of $2+5i$ and $-6-2i$ located?

Solution: Since $(2+5i)+(-6-2i)=-4+3i$, $a=-4$ and $b=3$. The ordered pair $(-4, 3)$ is located in **Quadrant II**. Notice in the diagram that the sum is the diagonal of the parallelogram whose adjacent sides are the directed segments from the origin to each of the complex ordered pairs.

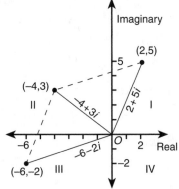

Simplifying Powers of *i*

Any nonnegative integer power of i can be reduced to either ± 1 or $\pm i$, as the following list suggests:

$i^0 = 1$	$i^4 = i^2 \cdot i^2 = 1$	$i^8 = i^4 \cdot i^4 = 1$
$i^1 = i$	$i^5 = i^4 \cdot i = i$	$i^9 = i^8 \cdot i = i$
$i^2 = -1$	$i^6 = i^4 \cdot i^2 = i^2 = -1$	$i^{10} = i^8 \cdot i^2 = i^2 = -1$
$i^3 = i^2 \cdot i = -i$	$i^7 = i^4 \cdot i^3 = i^3 = -i$	$i^{11} = i^8 \cdot i^3 = i^3 = -i$

Observe that consecutive integer powers of i greater than 3 follow a cyclic pattern that repeats every four integers. Since integer powers of i that are multiples of 4 are equal to 1, powers of i can be simplified by dividing the exponent by 4 and using the remainder as the new power of i. For example, i^{14} is equivalent to i^2, since $14 \div 4 = 3$, remainder 2.

Examples

6. Solve for x: $x^2 + 16 = 0$.

Solution: $x^2 = -16$
$$x = \pm\sqrt{-16}$$
$$= \pm 4i$$

7. Simplify:
(a) i^{52} **(b)** i^{25} **(c)** $i^7 \cdot i^{27}$

Solutions: To simplify i^n: (1) Divide n by 4 and note the remainder, R; then (2) write $i^n = i^R$.
(a) Divide: $52 \div 4 = 13$, remainder 0. Hence $i^{52} = i^0 = \mathbf{1}$.
(b) Divide: $25 \div 4 = 6$, remainder 1. Hence $i^{25} = i^1 = \mathbf{\textit{i}}$.
(c) $i^7 \cdot i^{27} = i^{34}$. Divide: $34 \div 4 = 8$, remainder 2. Hence $i^{34} = i^2 = \mathbf{-1}$.

Exercise Set 3.1

1–6. Determine whether each set of numbers forms a field under the operations of addition and multiplication. If the system does not form a field, state a field property the system lacks and illustrate, using specific numbers.

1. {rationals}   **2.** {irrationals}   **3.** {−1, 0, 1}
4. {pure imaginary numbers}
5. {all real numbers between −5 and 5, inclusive}
6. {all numbers of the form $q\sqrt{2}$, q is rational}

7 and 8. Subtract.

7. $3 - 3i$ from $-2 + 2i$ **8.** $-5 - 2i$ from $-8 + 2i$

9–14. Determine the sum and the quadrant in which it is located.

9. -4 and $1 + 5i$ **11.** $2 - 5i$ and $5 - 2i$ **13.** $5 - 2i$ and $-1 + 3i$
10. $-3i$ and $-2 - i$ **12.** $-3 - 9i$ and $1 + i$ **14.** $-7 + 3i$ and $8 - 8i$

15. Find the additive inverse of:
(a) $5 - 3i$ **(b)** $-1 - 2i$ **(c)** $3 + 7i$

16–24. Express in simplest form.

16. $\sqrt{-36}$ **17.** $2\sqrt{-16}$ **18.** $\sqrt{-28}$

19. $2\sqrt{-9} + 3\sqrt{-25}$ **20.** $3\sqrt{-1} - 2\sqrt{-4}$ **21.** $\sqrt{-12} + \sqrt{-27}$

22. $2\sqrt{-8} - 3\sqrt{-18}$ **23.** $-3\sqrt{-48} - 2\sqrt{-3}$ **24.** $2\sqrt{-50} - \sqrt{-2} + \sqrt{-8}$

25–27. Solve each equation for x. *Express complex roots in terms of* i.

25. $x^2 + 25 = 0$ **26.** $2x^2 + 3 = -5$ **27.** $3x^2 + 41 = 5$

28–32. Express in simplest form.

28. i^{60} **29.** i^{99} **30.** $i^{18} + i^0$ **31.** $i^3 \cdot i^{11}$ **32.** $(i^4)^6$

3.2 MULTIPLYING AND DIVIDING COMPLEX NUMBERS

KEY IDEAS

Complex numbers in $a + bi$ form may be multiplied and divided in much the same way that binomials that contain radicals are multiplied and divided. There is, however, one exception. If a and b are *both* negative numbers, then $\sqrt{a}\sqrt{b} \neq \sqrt{ab}$. For example, $\sqrt{-4}\sqrt{-9} \neq \sqrt{36}$. Rather,

$$\sqrt{-4}\sqrt{-9} = 2i \cdot 3i = 6i^2 = -6.$$

Multiplying Pure Imaginary Numbers

In general, if b and d are real numbers, then

$$bi \cdot di = (bd)(i^2) = bd(-1) = -bd.$$

Hence the product of two pure imaginary numbers is real, since i^2 can be replaced by -1.

Examples: 1. $2i \cdot 5i = 10i^2 = -10$

2. $\sqrt{-16}\sqrt{-9} = 4i \cdot 3i = 12i^2 = -12$

3. $(4i)^2 = 4i \cdot 4i = 16i^2 = -16$

4. $2i(3 - 5i) = 6i + (2i)(-5i)$

$$= 6i - 10i^2$$
$$= 6i - 10(-1)$$
$$= 6i + 10 \text{ or } 10 + 6i$$

Multiplying Complex Numbers

Complex numbers are treated as if they are binomials and are multiplied using FOIL.

$$\begin{array}{cccc} \underset{\vee}{F} & \underset{\vee}{O} & \underset{\vee}{I} & \underset{\vee}{L} \end{array}$$

Examples: 1. $(2+3i)(4+5i) = 2 \cdot 4 + (2 \cdot 5i + 3i \cdot 4) + 3i \cdot 5i$
$$\begin{aligned} &= \ 8 \ + (10i + 12i) \ + 15i^2 \\ &= \ 8 \ + \quad 22i \quad -15 \\ &= \ \textbf{--7 + 22}\boldsymbol{i} \end{aligned}$$

2. $(3+4i)(3-4i) = 3 \cdot 3 + [2(-4i) + 2(4i)] + (4i)(-4i)$
$$\begin{aligned} &= 3^2 + \quad\quad 0 \quad\quad + \ 4^2(-i^2) \\ &= 9 + \quad\quad\quad\quad\quad\quad 16 \\ &= \textbf{25} \end{aligned}$$

Notice in Example 2 that the product of $(3+4i)$ and $(3-4i)$ equals $3^2 + 4^2$ or 25. The product does not contain i. This happens whenever the two complex numbers form a *conjugate pair*. Two complex numbers are **complex conjugates** if they take the sum and difference of the same two terms. The complex conjugate of $a + bi$ is $a - bi$. The product of a pair of complex conjugates is *always* a positive real number.

MATH FACTS

MULTIPLYING COMPLEX CONJUGATES

$$(a + bi)(a - bi) = a^2 + b^2$$

Example

1. Express each product in simplest form:
(a) $(4+2i)(4-2i)$ **(b)** $(-3+i)(-3-i)$

Solutions: **(a)** Since $4 + 2i$ and $4 - 2i$ are complex conjugates, their product is $a^2 + b^2$, where $a = 4$ and $b = 2$:

$$(4+2i)(4-2i) = 4^2 + 2^2 = \textbf{20}.$$

(b) Since $-3 + i$ and $-3 - i$ are complex conjugates, their product is $a^2 + b^2$, where $a = -3$ and $b = 1$:

$$(-3+i)(-3-i) = (-3)^2 + (1)^2 = \textbf{10}.$$

Dividing Complex Numbers

If the denominator of a fraction is a complex number, then the quotient can be expressed in standard $a + bi$ form by multiplying the fraction by 1 in the form of the complex conjugate of the denominator divided by itself. For example,

$$\frac{3+i}{1-2i} = \frac{3+i}{1-2i}\left(\frac{1+2i}{1+2i}\right) = \frac{(3+1)(1+2i)}{1^2+2^2}$$

Use FOIL:

$$= \frac{3+(6i+i)+2i^2}{1+4}$$

Simplify:

$$= \frac{3+7i-2}{5}$$

Put into $a + bi$ form:

$$= \frac{1}{5} + \frac{7}{5}i$$

Example

2. Write the multiplicative inverse of $3 + 4i$ in standard form.

Solution: The multiplicative inverse of a complex number is its reciprocal. Thus, the multiplicative inverse of $3+4i$ is $\frac{1}{3+4i}$, which can be expressed in standard form by multiplying the numerator and the denominator by the complex conjugate of $3 + 4i$:

$$\frac{1}{3+4i}\left(\frac{3-4i}{3-4i}\right) = \frac{3-4i}{3^2+4^2} = \frac{3-4i}{9+16}$$

$$= \frac{3-4i}{25} = \frac{3}{25} - \frac{4}{25}i$$

Exercise Set 3.2

1–15. Express each product in simplest form.

1. $3i \cdot 5i$

2. $-2i \cdot 4i$

3. $(-5i)(-6i)$

4. $3i(-2+i)$

5. $-2i(-1-3i)$

6. $(3+2i)(1+i)$

7. $(2-i)(1-2i)$

8. $(-3+5i)(-2+i)$

9. $(1+3i)^2$

10. $(-2+4i)^2$

11. $(2-5i)(2+5i)$

12. $(-2+5i)(-2-5i)$

13. $(4-2i)(4+2i)$

14. $(\sqrt{3}-i)(\sqrt{3}+i)$

15. $(\sqrt{5}+i\sqrt{2})(\sqrt{5}-i\sqrt{2})$

16–20. Find each multiplicative inverse, and write it in standard $a + bi$ form.

16. $1-3i$ **17.** $2+7i$ **18.** $-3+5i$ **19.** $\sqrt{3}-i$ **20.** $\sqrt{7}+i\sqrt{2}$

21–26. Express each quotient in simplest a + bi *form.*

21. $\dfrac{2-5i}{i}$ **23.** $\dfrac{2-i}{2+i}$ **25.** $\dfrac{4+\sqrt{-9}}{4-\sqrt{-9}}$

22. $\dfrac{3i}{1-i}$ **24.** $\dfrac{3+2i}{1-2i}$ **26.** $\dfrac{\sqrt{5}-i}{\sqrt{5}}$

3.3   SOLVING QUADRATIC EQUATIONS BY FORMULA

⌃ KEY IDEAS ⌄

Starting with a linear equation of the form $ax + b = c$, you can solve for variable x in terms of constants a, b, and c:

$$ax = c - b, \quad \text{which means } x = \frac{c-b}{a} \quad (a \neq 0).$$

The resulting equation represents a general formula for solving linear equations having the form $ax + b = c$. For example, in the equation $2x + 3 = 11$, $a = 2$, $b = 3$, and $c = 11$, so that

$$x = \frac{c-b}{a} = \frac{11-3}{2} = \frac{8}{2} = 4.$$

Similarly, starting with a quadratic equation written in the standard form $ax^2 + bx + c = 0$ ($a \neq 0$), you can solve for variable x by using the method of completing the square. The resulting equation is called the **quadratic formula**.

Using the Quadratic Formula

Any quadratic equation (including equations that cannot be factored) that is put into the standard form $ax^2 + bx + c = 0$ can be solved by using the *quadratic formula*:

$$x = \frac{-b \pm \sqrt{b^2 - 4ac}}{2a} \quad (a \neq 0)$$

For example, to solve the equation $3x^2 + 2x = 1$, follow these steps:

Step	Example
1. Put the equation into standard form.	**1.** $3x^2 + 2x \;\; -1 = 0$ $\downarrow\;\;\;\;\; \downarrow\;\;\;\; \downarrow$
2. Identify the values for a, b, and c.	**2.** $ax^2 + bx + c = 0$ $a = 3,\ b = 2,\ \text{and } c = -1$
3. Write the quadratic formula, and replace the letters a, b, and c with their numerical values.	**3.** $x = \dfrac{-b \pm \sqrt{b^2 - 4ac}}{2a}$ $= \dfrac{-2 \pm \sqrt{(2)^2 - 4(3)(-1)}}{2(3)}$
4. Simplify	**4.** $x = \dfrac{-2 \pm \sqrt{4 + 12}}{6}$ $= \dfrac{-2 \pm \sqrt{16}}{6}$ $= \dfrac{-2 \pm 4}{6}$ $x = \dfrac{-2 + 4}{6} \quad \text{or} \quad x = \dfrac{-2 - 4}{6}$ $= \dfrac{2}{6} \qquad\qquad\quad = -\dfrac{6}{6}$ $= \dfrac{1}{3} \qquad\qquad\quad = -1$
5. Write the solution set.	**5.** $\left\{ \dfrac{1}{3}, -1 \right\}$

In the preceding example the roots were rational, indicating that the original equation was factorable. In the next example the equation is *not* factorable, so its roots are *not* rational.

Example

Solve by formula: $x^2 + 8 = 7x$.

Solution:

$$x^2 + 8 = 7x$$

Put the equation into standard form: $x^2 - 7x + 8 = 0$

Write the formula:

$$x = \frac{-b \pm \sqrt{b^2 - 4ac}}{2a}$$

Let $a = 1$, $b = -7$, and $c = 8$:

$$x = \frac{7 \pm \sqrt{(-7)^2 - 4(1)(8)}}{2(1)}$$

$$= \frac{7 \pm \sqrt{49 - 32}}{2}$$

$$= \frac{7 \pm \sqrt{17}}{2}$$

Write the solution set:

$$\left\{ \frac{7 + \sqrt{17}}{2}, \frac{7 - \sqrt{17}}{2} \right\}$$

Exercise Set 3.3

1. What are the roots of the quadratic equation $2x^2 - 5x + 1 = 0$?

(1) $\dfrac{5 \pm \sqrt{17}}{4}$ (2) $\dfrac{-5 \pm \sqrt{21}}{4}$ (3) $\dfrac{5 \pm \sqrt{33}}{-4}$ (4) $\dfrac{-5 \pm \sqrt{23}}{-4}$

2. What are the roots of the quadratic equation $2x^2 - x - 14 = 0$?

(1) $\dfrac{-1 \pm \sqrt{111}}{2}$ (2) $\dfrac{1 \pm \sqrt{111}}{4}$ (3) $\dfrac{1 \pm \sqrt{113}}{4}$ (4) $\dfrac{-1 \pm \sqrt{113}}{4}$

3–6. Solve algebraically for x, *and write the answer in simplest radical form.*

3. $2x + 4 = x^2$

4. $\dfrac{6}{x^2} - 3 = \dfrac{x + 4}{x}$

5. $\dfrac{1}{x} + \dfrac{1}{x+3} = 2$

6. $\dfrac{x+5}{2x-1} = \dfrac{2x+1}{x+3}$

7. A rectangle is said to have a golden ratio when $\dfrac{w}{h} = \dfrac{h}{w-h}$, where w represents width and h represents height. If $w = 3$, find h.

8. A homeowner wants to increase the size of a rectangular deck that now measures 15 feet by 20 feet, but building-code laws state that a deck cannot be larger than 900 square feet. If the length and the width are to be increased by the same amount, find, to the *nearest tenth*, the maximum number of feet by which the length of the deck may be legally increased.

3.4 SOLVING QUADRATIC EQUATIONS HAVING IMAGINARY ROOTS

KEY IDEAS

In the quadratic formula

$$X = \frac{-b \pm \sqrt{b^2 - 4ac}}{2a},$$

the radicand $b^2 - 4ac$ is called the **discriminant**. If the discriminant is negative, then the roots of the quadratic equation are imaginary and can be represented in terms of the imaginary unit, i.

The graph of a quadratic equation with imaginary roots is a parabola that has no x-intercepts.

Quadratic Equations and Parabolas

The graph of $y = ax^2 + bx + c$, where a, b, and c are real numbers and $a \neq 0$, is a parabola and may intersect the x-axis in two points, one point, or no point, depending on the nature of the roots of the related quadratic equation $ax^2 + bx + c = 0$. The possibilities are given in Table 3.2.

TABLE 3.2 RELATING THE PARABOLA $y = ax^2 + bx + c$ **TO THE NATURE OF THE ROOTS OF** $ax^2 + bx + c = 0$ $(a \neq 0)$

A Possible Graph of $y = ax^2 + bx + c$ $(a > 0)$	$ax^2 + bx + c = 0$ $(a \neq 0)$	
	Discriminant	**Nature of Roots**
	$b^2 - 4ac > 0$ (Two x-intercepts)	Real and unequal. If $b^2 - 4ac$ is a perfect square, roots are rational; otherwise, roots are irrational.
	$b^2 - 4ac = 0$ (One x-intercept, so graph is tangent to the x-axis.)	Real, rational, and equal.
	$b^2 - 4ac < 0$ (No x-intercept)	Imaginary and unequal.

Solving Quadratic Equations

Imaginary roots of quadratic equations arise when the discriminant is less than 0 and always occur in conjugate pairs. If $a + bi$ is a root of a quadratic equation, then $a - bi$ must be the other root.

Example

1. Express the roots of $x^2 + 5 = 2x$ in $a + bi$ form.

Solution: Put the equation into standard form:

$$x^2 - 2x + 5 = 0.$$

Write the quadratic formula: $x = \dfrac{-b \pm \sqrt{b^2 - 4ac}}{2a}$

Let $a = 1$, $b = -2$, $c = 5$: $x = \dfrac{-(-2) \pm \sqrt{(-2)^2 - 4(1)(5)}}{2(1)}$

$$= \frac{2 \pm \sqrt{4 - 20}}{2}$$

$$= \frac{2 \pm \sqrt{-16}}{2}$$

$$= \frac{2 \pm 4i}{2}$$

$$= \frac{2}{2} \pm \frac{4i}{2} = 1 \pm 2i$$

The roots of $x^2 + 5 = 2x$ are **$1 + 2i$** and **$1 - 2i$**.

Describing the Type of Roots

It is not necessary to solve a quadratic equation in order to determine the *nature* of its roots. The value of the discriminant tells us the type of roots (see Table 3.2).

Example

2. Find the discriminant and describe the nature of the roots:
(a) $3x^2 - 11x = 4$ **(b)** $-2x^2 + 3x - 4 = 0$

Solutions: **(a)** Write the equation in standard form:

$$3x^2 - 11x - 4 = 0.$$

Let $a = 3$, $b = -11$, and $c = -4$. Next, evaluate the discriminant:

$$b^2 - 4ac = (-11)^2 - 4(3)(-4)$$
$$= 121 + 48 = \mathbf{169}$$

Since the discriminant is a perfect square ($169 = 13 \times 13$), the roots of $3x^2 - 11x = 4$ are **real**, **rational**, and **unequal**.

(b) Since $-2x^2 + 3x - 4 = 0$, let $a = -2$, $b = 3$, and $c = -4$. Then

$$b^2 - 4ac = 3^2 - 4(-2)(-4)$$
$$= 9 - 32 = -23$$

Since the discriminant is less than 0, the roots of $-2x^2 + 3x - 4 = 0$ are **imaginary** and, as a result, unequal.

Finding an Unknown Coefficient (*a*, *b*, or *c*)

The discriminant can be used to determine the value that an unknown coefficient in a quadratic equation must have in order for the roots to be of a given type.

Examples

3. Find the *smallest* integer value of k that makes the roots of the equation $x^2 - 5x + k = 0$ imaginary.

Solution: For the equation $x^2 - 5x + k = 0$, let $a = 1$, $b = -5$, and $c = k$. If the roots are imaginary, the discriminant is less than 0.

$$b^2 - 4ac < 0$$
$$(-5)^2 - 4(1)(k) < 0$$
$$25 - 4k < 0$$
$$-4k < -25$$

Divide by -4 and reverse the direction of the inequality:

$$\frac{-4k}{-4} > \frac{-25}{-4}$$
$$k > 6\frac{1}{4}$$

The *smallest* integer value of k that makes the inequality a true statement is 7. Hence $k = \mathbf{7}$.

4. For what value of k is the graph of $y = kx^2 - 12x + 9$ tangent to the x-axis?

Solution: A parabola is tangent to the x-axis if it intersects the x-axis in exactly one point. This situation corresponds to the one in which the roots of the related quadratic equation $kx^2 - 12x + 9 = 0$ are real and *equal*. Let $a = k$, $b = -12$, and $c = 9$. If the roots are real and equal, the value of the discriminant must be 0.

$$b^2 - 4ac = -0$$
$$(-12)^2 - 4k(9) = 0$$
$$144 - 36k = 0$$
$$-36k = -144$$
$$k = \frac{-144}{-36} = \mathbf{4}$$

Exercise Set 3.4

1–9. Solve each equation, and express its roots in a + bi *form.*

1. $x^2 + 13 = 4x$ **4.** $-x^2 = 4(2x+5)$ **7.** $3(x^2+1) = 5x$

2. $x^2 = 6x - 25$ **5.** $-x^2 + 2x = 10$ **8.** $2x^2 + 5 = 6x$

3. $\dfrac{x^2}{2} = x - 1$ **6.** $x + \dfrac{5}{x} = 2$ **9.** $2(x+2) = -\dfrac{5}{x}$

10. For the graph of $y = x^2 - 2kx + 16$, determine the value of k such that:
    (a) the graph is tangent to the x-axis and k is positive.
    (b) k is the largest integer for which the graph has no x-intercept.

11–14. For each equation find the discriminant and describe the nature of the roots.

11. $3x^2 + 5x + 2 = 0$ **13.** $-2x^2 + 3x = 2$

12. $2x^2 - 1 = 5x$ **14.** $x - 1 = x^2 - 2x + 1$

15–18. For each equation find the value or values of k *that make the roots real and equal.*

15. $x^2 + kx + 25 = 0$ **17.** $kx^2 - 14x + 49 = 0$

16. $9x^2 + 6x + k = 0$ **18.** $kx^2 - 11kx + 121 = 0$

19–22. For each equation find the largest integral value of k *that makes the roots real.*

19. $x^2 - 4x + k = 0$ **21.** $2x^2 - 7x + k = 0$

20. $kx^2 - 9x + 3 = 0$ **22.** $kx^2 - 10x + k = 0$

23. What is the *smallest* integral value of k that makes the roots of the equation $2x^2 - 5x + k = 0$ imaginary?

3.5 APPLYING THE SUM AND THE PRODUCT OF THE ROOTS FORMULAS

━━━━━━━━━━━━━━━━━━━ **KEY IDEAS** ━━━━━━━━━━━━━━━━━━━

If $ax^2 + bx + c = 0$ $(a \neq 0)$, then the sum and the product of the roots may be determined without solving the equation by using these formulas:

$$\text{Sum of roots} = -\frac{b}{a} \quad \text{and} \quad \text{Product of roots} = \frac{c}{a}.$$

A quadratic equation can be formed when only its roots are known by writing

$$x^2 - (\text{sum of roots})\,x + (\text{product of roots}) = 0.$$

Checking Roots of Quadratic Equations

Rather than substituting into the original equation, the roots obtained by solving a quadratic equation of the form $ax^2 + bx + c = 0$ $(a \neq 0)$ can be checked by verifying that their sum equals $-\frac{b}{a}$ and their product equals $\frac{c}{a}$.

Example

1. Determine whether $3 \pm \sqrt{10}$ are the roots of $x^2 - 6x = 1$.

Solution: If $x^2 - 6x = 1$, then $x^2 - 6x - 1 = 0$, where $a = 1$, $b = -6$, and $c = -1$.

$$\text{Sum} = -\frac{b}{a} = -\frac{(-6)}{1} = 6 \quad \text{and} \quad \text{Product} = \frac{c}{a} = \frac{-1}{1} = -1.$$

The roots to be checked are $3 - \sqrt{10}$ and $3 + \sqrt{10}$. If their sum is 6 and their product is -1, then these must be the roots of the given equation.

$$\text{Sum} = (3 - \sqrt{10}) + (3 + \sqrt{10}) \quad \text{and} \quad \text{Product} = (3 - \sqrt{10})(3 + \sqrt{10})$$
$$= 6 \qquad\qquad\qquad\qquad\qquad = (3)^2 - (\sqrt{10})^2$$
$$= 9 - 10$$
$$= -1$$

Since the sums agree and the products agree, $\mathbf{3 \pm \sqrt{10}}$ **are the roots** of $x^2 - 6x = 1$.

Forming Quadratic Equations Whose Roots Are Given

A quadratic equation having a given pair of numbers r_1 and r_2 as its roots can be formed by writing

$$x^2 - (r_1 + r_2)x + r_1 \cdot r_2 = 0.$$

Keep in mind that irrational and imaginary roots must occur in conjugate pairs.

Examples

2. Find a quadratic equation that has $5 + 2i$ as one of its roots.

Solution: Imaginary roots always occur in conjugate pairs. Let $r_1 = 5 + 2i$ and $r_2 = 5 - 2i$. Then

$$\text{Sum} = r_1 + r_2 = (5 + 2i) + (5 - 2i) = 10$$
$$\text{Product} = r_1 r_2 = (5 + 2i)(5 - 2i)$$
$$= 5^2 + 2^2 = 29$$

In general, $x^2 - (r_1 + r_2)x + r_1 r_2 = 0$. Substituting for the sum and product gives $x^2 - 10x + 29 = 0$.

3. The roots of the equation $x^2 + px + q = 0$ are $1 \pm \sqrt{3}i$. Find the values of: **(a)** p **(b)** q

Solutions: The value of p equals the negative of the sum of the roots, and the value of q equals the product of the roots. Let $r_1 = 1 + \sqrt{3}i$ and $r_2 = 1 - \sqrt{3}i$.

(a) $p = -(r_1 + r_2) = -[(1 + \sqrt{3}i) + (1 - \sqrt{3}i)]$
$$= -(2 + 0) = -2$$

(b) $q = r_1 \cdot r_2 = (1 + \sqrt{3}i)(1 - \sqrt{3}i)$
$$= 1^2 + (\sqrt{3})^2$$
$$= 1 + 3 = 4$$

Exercise Set 3.5

1. Given the equation $x^2 - x + q = 0$, one of whose roots is 3.
   (a) Find the sum of the roots.
   (b) Using your answer to part **(a)**, find the value of q.
   (c) Find the value of q by replacing x by 3 in the original equation.

2. One root of the equation $6x^2 + kx + 5 = 0$ is 1.
   (a) What is the other root?   **(b)** What is the value of k?

3. By what amount does the sum of the roots of the equation $3x^2-6x+3=0$ exceed the product of the roots?

4 and 5. Given the equation $x^2+px+q=0$.

4. If the roots are $\pm i$, find the values of p and q.
5. If the roots are $2 \pm \sqrt{6}$, find the values of p and q.

6. In the equation $ax^2+bx+a=0$, a and b are integers. If one root is $\frac{1}{2}$, find:
   (a) the other root   **(b)** the sum and product of the roots

7–10. Write a quadratic equation with integer coefficients whose roots are given.

7. $\{1, -2\}$ **9.** $\left\{\dfrac{1}{4}, -\dfrac{1}{2}\right\}$

8. $\{2i, -2i\}$ **10.** $\{1 \pm \sqrt{7}\}$

REGENTS TUNE-UP: CHAPTER 3

Each of the questions in this section has appeared on a previous Course III Regents Examination. Here is an opportunity for you to review Chapter 3 and, at the same time, prepare for the Course III Regents Examination.

---

1. Express $3i(1-i)$ in $a + bi$ form.

2. Express the product $(3-i)(4+i)$ in the form $a + bi$.

3. Expressed in simplest form, $2\sqrt{-50} - 3\sqrt{-8}$ is equivalent to
   (1) $16i\sqrt{2}$ (2) $3i\sqrt{2}$ (3) $4i\sqrt{2}$ (4) $-\sqrt{-42}$

4. The roots of the equation $x^2 + x + 1 = 0$ are
   (1) real, rational, and unequal   (3) real, rational, and equal
   (2) real, irrational, and unequal   (4) imaginary

5. What is the sum of the roots of the equation $3x^2 = 9x + 1$?

6. Solve the equation $3x^2 - 4x = -2$, and express the roots in $a + bi$ form.

7. If the equation $9x^2 - 12x + k = 0$ has equal roots, find the value of k.

8. Express the roots of the equation $3x^2 = -2(2x+3)$ in $a + bi$ form.

9. If the sum of the roots of the equation $2x^2 - 5x - 3 = 0$ is added to the product of the roots, the result is
   (1) 1   (2) $-\dfrac{1}{4}$ (3) -1 (4) 4

10. Solve the equation

$$\frac{x+8}{5} + \frac{x+5}{x} = 1,$$

and express the roots in the form $a + bi$.

11. What is the product of the roots of the equation $2x^2 - 9x + 6 = 0$?
   (1) $\dfrac{9}{2}$ (2) $-\dfrac{9}{2}$ (3) 3   (4) $\dfrac{1}{3}$

12. The expression $\dfrac{1}{5+2i}$ is equivalent to

(1) $\dfrac{5+2i}{21}$ (3) $\dfrac{5-2i}{21}$

(2) $\dfrac{5+2i}{29}$ (4) $\dfrac{5-2i}{29}$

13. If $4 + 2i - (a + 4i) = 9 - 2i$, find the value of a.

14. If $Z_1 = -1 + 6i$ and $Z_2 = 4 + 2i$, graphically represent Z_1, Z_2, and $Z_1 + Z_2$.

ANSWERS TO SELECTED EXERCISES: CHAPTER 3

Section 3.1
1. System is a field.
3. Since $1 + 1 = 2$ and 2 is not an element of $\{-1, 0, 1\}$, the system is not closed under addition, so it does not form a field.
5. Since $-5 \times 5 = -25$ and -25 is not an element of the set of all real numbers from -5 to 5, inclusive, the system is not closed under multiplication, so it does not form a field.

7. $-5 + 5i$ **15. (a)** $-5 + 3i$ **25.** $x = \pm 5i$
9. $-3 + 5i$, Quadrant II **(b)** $1 + 2i$ **(c)** $-3 - 7i$ **27.** $x = \pm 2i\sqrt{3}$
10. $-2 - 4i$, Quadrant III **17.** $8i$ **28.** 1
11. $7 - 7i$, Quadrant IV **19.** $21i$ **29.** $-i$
13. $4 + i$, Quadrant I **21.** $5i\sqrt{3}$ **30.** 0
14. $1 - 5i$, Quadrant IV **22.** $-5i\sqrt{2}$ **31.** -1
  23. $-14i\sqrt{3}$ **32.** 1

Section 3.2
1. -15 **9.** $-8 + 6i$ **17.** $\dfrac{2}{53} - \dfrac{7}{53}i$ **23.** $\dfrac{3}{5} - \dfrac{4}{5}i$
3. -30 **11.** 29
5. $-6 + 2i$ **13.** 20 **19.** $\dfrac{\sqrt{3}}{4} + \dfrac{1}{4}i$ **25.** $\dfrac{7}{25} + \dfrac{24}{25}i$
7. $-5i$ **15.** 7 **21.** $-5 - 2i$

Section 3.3
1. (1) **3.** $1 \pm \sqrt{5}$ **5.** $\dfrac{-2 \pm \sqrt{10}}{2}$ **7.** $\dfrac{-3 \pm 3\sqrt{5}}{2}$

2. (3) **4.** $\dfrac{-1 \pm \sqrt{7}}{2}$ **6.** $\dfrac{4 \pm 4\sqrt{2}}{3}$ **8.** 12.6

Section 3.4

1. $2 \pm 3i$

2. $3 \pm 4i$

3. $1 \pm i$

4. $-4 \pm 2i$

5. $1 \pm 3i$

6. $1 \pm 2i$

7. $\dfrac{5}{6} \pm \dfrac{\sqrt{11}}{6}\, i$

8. $\dfrac{3}{2} \pm \dfrac{1}{2}\, i$

9. $-1 \pm \dfrac{\sqrt{6}}{2}\, i$

10. (a) 4 **(b)** 3

11. Discriminant is 1, so roots are real, rational, and unequal.

13. Discriminant is -7, so roots are imaginary.

15. ± 10

17. 1

19. 3

21. 6

23. 4

Section 3.5

1. (a) -2 **(b)** -6

2. (a) $\frac{5}{6}$ **(b)** -11

3. 1

4. $p = 0,\ q = 1$

5. $p = -4,\ q = -2$

6. (a) 2

(b) Sum $= \dfrac{5}{2}$, product $= 1$

7. $x^2 + x - 2 = 0$

8. $x^2 + 4 = 0$

9. $8x^2 + 2x - 1 = 0$

10. $x^2 - 2x - 6 = 0$

Regents Tune-Up: Chapter 3

1. $3 + 3i$

2. $13 - i$

3. (3)

4. (4)

5. 3

6. $\dfrac{2}{3} \pm \dfrac{\sqrt{2}}{3}\, i$

7. 4

8. $-\dfrac{2}{3} \pm \dfrac{\sqrt{14}}{3}\, i$

9. (1)

10. $-4 \pm 3i$

11. (3)

12. (4)

13. -5

14.

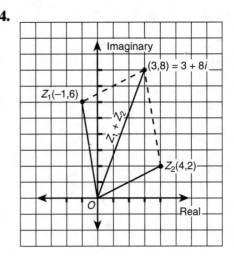

FUNCTIONS AND CIRCLES

CHAPTER **4**

RELATIONS, FUNCTIONS, AND TRANSFORMATIONS

4.1 DEFINING RELATIONS AND FUNCTIONS

KEY IDEAS

A **function** in two variables describes how one quantity depends on another quantity. A function may take the form of a set of ordered pairs, an equation in two variables, or a graph. Regardless of the form it takes, a function must obey the condition that no two of its ordered pairs have the same first member (x) but different second members (y).

Definition of Relations

A **relation** is a set of ordered pairs. The set of all first members of the ordered pairs is called the **domain**. The set of all second members of the ordered pairs is called the **range**. For example, here is a relation that pairs three pupils with their heights:

{(John, 70 inches),　(Maria, 68 inches),　(George, 66 inches)}.

The domain and range of this relation are as follows:

Domain = {John, Maria, George}
Range = {70, 68, 66}

Definition of Function

A function is a special type of relation. A **function** is a set of ordered pairs of the form (x, y) in which no two ordered pairs have the same value of x but different values of y. The ordered pairs that comprise a function may appear as a list, an equation, or a graph. If

f = {(−1, 1), (0, 0), (1, 1), (2, 4)},

65

then relation f is a function. However, if

$$g = \{(1, 2), (2, 1), (1, 3)\},$$

then relation g is *not* a function since the ordered pairs (1, 2) and (1, 3) have the same value of x but different values of y.

Describing Functions Using Equations

Some functions consist of a set of ordered pairs in which the values of y can be obtained from the corresponding values of x by using an equation. For example, the equation $y = x + 1$ defines a function consisting of the set of all ordered pairs whose second member, y, is one more than its first member, x. Notice that the value of y *depends* on the value of x. If $x = 3$, then $y = 4$; if $x = 7.5$, then $y = 8.5$; and so forth. For this reason, the variable y is sometimes referred to as the **dependent** variable, while the variable x is called the **independent** variable.

An equation that describes a function may be thought of as a *rule* that tells how the elements of the domain are paired or assigned to the elements of the range. For example, suppose that function g is described by the equation $y = 3x$. If the domain of g is $\{1, 2, 3\}$, then the function rule $y = 3x$ tells us to pair the numbers 1, 2, and 3 with numbers that are three times as great. Thus, $g = \{(1, 3), (2, 6), (3, 9)\}$.

Example

1. Determine whether each set of ordered pairs or sentences describes a function:

(a) $\{(4, 4), (4, 0), (2, 2)\}$ (c) $y = x^3$ (e) $y = |x|$

(b) $\{(1, 5), (0, 5), (2, 3)\}$ (d) $y > x$ (f) $x = y^2$

Solutions: **(a)** The set of ordered pairs is *not* a function since two ordered pairs, (4, 4) and (4, 0), have the same first member but different second members.

(b) The set of ordered pairs is a function.

(c) For each value of x, the equation produces exactly one value of y. Hence, the equation $y = x^3$ describes a function.

(d) The relation $y > x$ is *not* a function since it includes ordered pairs such as (1, 2) and (1, 3) that have the same value of x paired with different values of y.

(e) The equation $y = |x|$ describes a function since each value of x (positive, negative, or zero) determines exactly one value of y.

(f) The equation $x = y^2$ does *not* describe a function since ordered pairs such as (4, 2) and (4, –2) satisfy the equation but have different values of y for the same value of x.

Describing a Function as a Mapping

In Figure 4.1, the elements −1, 0, 1, and 2 of the domain of function f are paired with or "mapped onto" the corresponding elements of the range, using the function rule $y = x^2$. Since each element x of the domain is paired with x^2 in the range, the *mapping* that describes this function may be represented in symbols as **f: $x \to x^2$**.

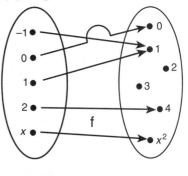

DOMAIN   **RANGE**

Figure 4.1 Domain and Range

Function Notation

Functions are usually designated by lower-case letters such as f and g. For any function f the value of y that corresponds to a *particular* value of x is denoted by f(x). If function f is described by the equation $y = 4x$, then the notation f(2) represents the value of y when $x = 2$. If $x = 2$, then $y = 4 \cdot 2 = 8$. Hence f(2) = 8. The expression f(2) is read as "f of 2." Similarly, f(3) = 12 since, when $x = 3$, $y = 4 \cdot 3 = 12$.

Since y and f(x) mean the same thing, $y = $ f(x). The equation $y = 4x$ may be written as f(x) = 4x.

Examples

2. If $y = x^2 + 1$, find f(5).

Solution: The notation f(5) tells you to find the value of y when $x = 5$.

Let $y = $ f(x): $y = $ f(x) $= x^2 + 1$
Replace x by 5:   f(5) $= 5^2 + 1$
Simplify: **= 26**

3. If f(x) $= x^{\frac{3}{2}} - (x-1)^0$, find f(9).

Solution: Replace x by 9 and simplify:

$$\begin{aligned}
\text{f}(9) &= 9^{\frac{3}{2}} & -(9-1)^0 \\
&= (\sqrt{9})^3 & -8^0 \\
&= 3^3 & -1 \\
&= 27 & -1 \\
&= \mathbf{26}
\end{aligned}$$

67

4. If g = {(–2, 5), (–1, 2), (0, 1), (1, 2), (2, 5)}, find g(1).

Solution: The notation g(1) tells you to find the value of *y* in the ordered pair in which *x* = 1. In the ordered pair (1, 2), *x* = 1 and *y* = 2. Hence g(1) = **2**.

5. If $g(t) = t^2$, find:
(a) 3g(2) **(b)** $[g(-3)]^2$ **(c)** g(4*n*) **(d)** g(*x* + 1)

Solutions:
(a) Let *t* = 2. Since $g(2) = 2^2 = 4$, then $3[g(2)] = 3[4] = $ **12**.
(b) Let *t* = –3. Since $g(-3) = (-3)^2 = 9$, then $[g(-3)]^2 = [9]^2 = $ **81**.
(c) Let *t* = 4*n* and simplify.   **(d)** Let *t* = *x* + 1 and simplify.

$$g(t) = t^2$$
$$g(4n) = (4n)^2$$
$$= \mathbf{16n^2}$$

$$g(t) = t^2$$
$$g(x + 1) = (x + 1)^2$$
$$= \mathbf{x^2 + 2x + 1}$$

Restricting the Domain and Range

Unless otherwise indicated, the domain and range of a function are the largest possible sets of real numbers. In order that a real-valued function be defined for each number in its domain, it may be necessary to exclude one or more real numbers from the domain. For example, if $y = \frac{x}{x-5}$, then *x* cannot equal 5 since this value makes the denominator 0. Hence the domain of this function consists of all real numbers *except* 5.

To illustrate further, if $y = \sqrt{x-1}$, then *x* must take on values greater than or equal to 1 in order for the radicand to remain nonnegative. Hence the domain of this function consists of all real numbers greater than or equal to 1. Since *y* is equal to the square root of a nonnegative number, *y* is also nonnegative. The domain and range of this function can be expressed using set notation as follows:

$$\text{Domain} = \{x: x \geq 1\}$$
$$\text{Range} = \{y: y \geq 0\}$$

Exercise Set 4.1

1–8. For each relation, find the domain and the range.

1. {(–2, 2), (3, 2), (0, 3)}

2. {(–4, 0), (–1, 0), (1, 1)}

3. $y = 1 - x$

4. $y = \sqrt{x+3}$

5. $y = \sqrt{x-3}$

6. $y = \sqrt{3-x}$

7. $y = \sqrt{x+3}$

8. $y = \dfrac{x}{x^2 - 1}$

9. What is the domain of $h(x) = \sqrt{x^2 - 4x - 5}$?

(1) $\{x \mid x \geq 1 \text{ or } x \leq 5\}$ (3) $\{x \mid -1 \leq x \leq 5\}$

(2) $\{x \mid x \geq 5 \text{ or } x \leq -1\}$ (4) $\{x \mid -5 \leq x \leq 1\}$

10. Which of the following relations is *not* a function?

(1) $x = |y|$ (2) $2x + 3y = 6$ (3) $x = \sqrt{y}$ (4) $x = y^3$

11. For each function, express f(4) in simplest form.

(a) $f(x) = x^2 - 4x + 1$ (d) $f(t) = (2t)^{\frac{2}{3}}$

(b) $f(x) = x^2 - x^{\frac{1}{2}}$ (e) $f(x) = \sqrt{x^2 + 9} + \sqrt{9x}$

(c) $f(s) = (2s)^0 + 2s^0 + 2^s$ (f) $y = x^{\frac{3}{2}} + x^{-\frac{1}{2}}$

12. If $s = \{(-2, 4), (-1, 1), (0, 6), (1, -1), (2, 8)\}$, find:

(a) $s(-2) + s(2)$ (b) $3s(0)$ (c) $[s(2)]^{\frac{1}{3}}$

13. If $f(x) = x + 2$ and $g(x) = x^2 + 2x - 4$, for which values of x is $f(x) = g(x)$?

14. If $h(t) = (t + 1)^2$, find:

(a) h(−1)   (b) $[h(3)]^2$ (c) 2h(−4)   (d) h(2y)   (e) h(x − 1)

4.2   RECOGNIZING WHEN A GRAPH IS A FUNCTION

KEY IDEAS

Since a graph consists of a set of ordered pairs, it describes a relation that may or may not be a function.

Vertical-Line Test

In Figure 4.2, a vertical line intersects the graph in more than one point. The graph contains two points that have the same x-value but different y-values. Thus, the graph does not represent a function.

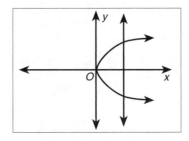

Figure 4.2   Vertical-Line Test

VERTICAL-LINE TEST

A graph represents a function if *no* vertical line intersects it in more than one point. The graph at the right represents a function since it passes the vertical-line test.

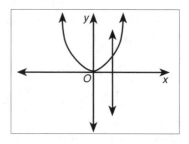

Recognizing Graphs of Quadratic Relations

The graphs of equations of the form

$$ax^2 + by^2 = c \quad (a, b, \text{ and } c \neq 0)$$

are given special names, which are summarized in Table 4.1. In each case the graphs fail the vertical-line test, so they do not represent functions.

Example

Name the graph of each equation:
(a) $4y^2 = 3 + 4x^2$ **(b)** $3x^2 = 18 - 3y^2$ **(c)** $-3x^2 - 5y^2 + 15 = 0$

Solutions: Rewrite each equation in the form $ax^2 + by^2 = c$.
(a) If $4y^2 = 3 + 4x^2$, then $-4x^2 + 4y^2 = 3$. This equation corresponds to the general form $ax^2 + by^2 = c$, where a and b have opposite signs. Hence the graph is a **hyperbola**.
(b) If $3x^2 = 18 - 3y^2$, then $3x^2 + 3y^2 = 18$. This equation corresponds to the general form $ax^2 + by^2 = c$, where a and b are equal and a, b, and c have the same sign. Hence the graph is a **circle**. Notice that dividing each member of the equation by 3 gives $x^2 + y^2 = 6$. Hence the radius of this circle is $\sqrt{6}$.
(c) If $-3x^2 - 5y^2 + 15 = 0$, then $15 = 3x^2 + 5y^2$, which also can be written as $3x^2 + 5y^2 = 15$. This equation corresponds to the general form $ax^2 + by^2 = c$, where $a \neq b$ and a, b, and c have the same sign. Hence the graph is an **ellipse**.

TABLE 4.1 EQUATIONS OF GRAPHS OF QUADRATIC RELATIONS

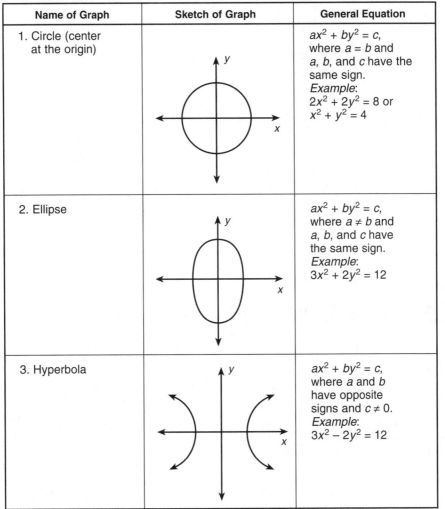

Name of Graph	Sketch of Graph	General Equation
1. Circle (center at the origin)		$ax^2 + by^2 = c$, where $a = b$ and a, b, and c have the same sign. *Example*: $2x^2 + 2y^2 = 8$ or $x^2 + y^2 = 4$
2. Ellipse		$ax^2 + by^2 = c$, where $a \neq b$ and a, b, and c have the same sign. *Example*: $3x^2 + 2y^2 = 12$
3. Hyperbola		$ax^2 + by^2 = c$, where a and b have opposite signs and $c \neq 0$. *Example*: $3x^2 - 2y^2 = 12$

Exercise Set 4.2

1–9. Name the graph of each relation, and state whether the graph represents a function.

1. $3x^2 = 12 - 3y^2$

2. $4x^2 + 8 = 2y^2$

3. $18 = 3y^2 + 3x^2$

4. $2x^2 - 13 = y$

5. $2x + 2y = 32$

6. $5y^2 = 3 - x^2$

7. $15 - 4x^2 = 4y^2$

8. $x^2 = 2y$

9. $0 = x^2 - (y^2 + 10)$

10–13. State whether each graph represents a function.

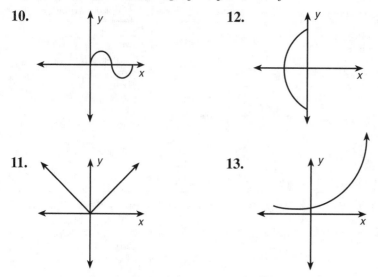

10.

12.

11.

13.

4.3 COMPOSING TWO FUNCTIONS

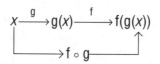

KEY IDEAS

Functions f and g may be linked together to form a *composite function* by using the output of function g as the input of function f:

$$x \xrightarrow{\ g\ } g(x) \xrightarrow{\ f\ } f(g(x))$$
$$\llcorner\!\!\longrightarrow f \circ g \longrightarrow\!\!\lrcorner$$

The new function is called the **composition** of function g *followed by* function f and is denoted by **f ∘ g**. The function value for this composite function is represented by (f ∘ g) (x) or, equivalently, by f(g(x)).

The composite function that puts the output of function f into function g is called the **composition** of function f *followed by* function g and is denoted by **g ∘ f**. The function value for this composite function is represented by (g ∘ f)(x) or, equivalently, by g(f(x)).

Evaluating Composite Functions

If $f(x) = x^2$ and $g(x) = 3x$, then f(g(2)) is determined by evaluating function g at $x = 2$ and putting the result into function f. Since $g(2) = 3 \cdot 2 = 6$,

$$f(g(2)) = f(6) = 6^2 = \mathbf{36}.$$

If $x = 2$, $g(f(2))$ is determined by evaluating function f at $x = 2$ and putting the result into function g. Since $f(2) = 2^2 = 4$,

$$g(f(2)) = g(4) = 3 \cdot 4 = \mathbf{12}.$$

Notice that $f(g(2))$ and $g(f(2))$ are not equal. *The order in which two functions are composed matters.*

Examples

1. If $f(x) = 2x + 1$ and $g(x) = |x|$, find:
(a) $f(g(-3))$ **(b)** $(g \circ f)(-3)$

Solutions: **(a)** Since $g(-3) = |-3| = 3$,

$$f(g(-3)) = f(3) = 2 \cdot 3 + 1 = \mathbf{7}$$

(b) The expression $(g \circ f)(-3)$ means $g(f(-3))$. If $f(x) = 2x + 1$, then $f(-3) = 2(-3) + 1 = -5$. Since $g(x) = |x|$,

$$g(f(-3)) = g(-5) = |-5| = \mathbf{5}$$

2. Let $f = \{(1, 2), (3, 8), (5, 4), (7, 8)\}$, and $g = \{(1, 3), (2, 6), (8, 5)\}$. Find:
(a) $f(g(1))$ **(b)** $g(f(7))$ **(c)** $f(g(2))$ **(d)** $g(f(5))$

Solutions: Evaluate by starting with the "inside" function.
(a) Since function g includes the ordered pair $(1, 3)$, then $g(1) = 3$, so $f(g(1)) = f(3)$. Function f includes the ordered pair $(3, 8)$, so $f(3) = 8$. Hence $f(g(1)) = \mathbf{8}$.
(b) $g(f(7)) = g(8) = \mathbf{5}$.
(c) $f(g(2)) = f(6)$. Since 6 is not in the domain of function f, $f(g(2))$ is not defined. In general, in order for $f(g(x))$ to be defined, $g(x)$ must be in the domain of function f.
(d) $g(f(5)) = g(4)$. Since 4 is not in the domain of function g, $g(f(5))$ is not defined. In general, in order for $g(f(x))$ to be defined, $f(x)$ must be in the domain of function g.

Finding an Equation for a Composite Function

If $f(x) = x^2$ and $g(x) = 3x$, then a single equation that defines $f(g(x))$ can be determined as follows:

Write the "outside" function: $\qquad$ $f(x) = [x]^2$
Replace x by the "inside" function: $\qquad$ $f(g(x)) = [g(x)]^2$
Express the right side in terms of x: $\qquad$ $= [3x]^2$
Simplify: $\qquad$ $= 9x^2$

To find an equation for $g(f(x))$, proceed as follows:

Write the "outside" function: $\qquad$ $g(x) = 3[x]$
Replace x by the "inside" function: $\qquad$ $g(f(x)) = 3[f(x)]$
Express the right side in terms of x: $\qquad$ $= 3[x^2]$
Simplify: $\qquad$ $= 3x^2$

Once again, notice that the composition of two functions is *not* a commutative operation since $f(g(x))$ and $g(f(x))$ produce different equations.

Examples

3. Let $f(x) = x^2 + 1$ and $g(x) = \sqrt{x}$, where $x \geq 0$. Find:
(a) $g(f(x))$ **(b)** $f(g(x))$

Solutions: **(a)** $\quad g(x) = \sqrt{x}$
$$g(f(x)) = \sqrt{[f(x)]}$$
$$= \sqrt{x^2 + 1}$$

(b) $\quad f(x) = x^2 + 1$
$$f(g(x)) = [g(x)]^2 + 1$$
$$= [\sqrt{x}]^2 + 1$$
$$= x + 1$$

4. If $f(x) = 2x$, find $f(f(x))$.

Solution: $f(f(x)) = f(2x) = 2[2x] = 4x$.

Exercise Set 4.3

1–6. For the given definitions of functions f and g, find:

(a) $f(g(x))$ **(b)** $g(f(x))$ **(c)** $f(f(x))$ **(d)** $g(g(x))$

1. $f(x) = x^2; g(x) = \dfrac{x}{2}$

2. $f(x) = 2x + 3; g(x) = 3x + 2$

3. $f(x) = \sqrt{x}; g(x) = x^2$

4. $f(x) = x + 1; g(x) = \dfrac{1}{x}$ $(x \neq 0)$

5. $f(x) = x^2; g(x) = 1 - x$

6. $f(x) = 4x - 1; g(x) = \dfrac{x+1}{4}$

7. If $f(x) = x^3$, find:

(a) $f(f(2))$ **(b)** $f\left(f\left(\dfrac{x}{2}\right)\right)$ **(c)** $f(f(a^2))$.

8. If $f(x) = 2x$ and $g(x) = x^2 + 8$, find all values of x such that $f(g(x)) = g(f(x))$.

9. If $f(x) = 2x + k$ and $g(x) = \dfrac{x-5}{2}$, for what value of k are $f(g(x))$ and $g(f(x))$ always equal?

10. If $f = \{(1, -1), (4, 6), (6, 1)\}$ and $g = \{(-1, 6), (4, 1), (1, 3)\}$, evaluate each of the following, provided that it exists:
(a) $f(g(4))$　　**(c)** $g(f(6))$　　**(e)** $(f \circ g)(4)$
(b) $g(g(4))$　　**(d)** $(g \circ f)(4)$　　**(f)** $(f \circ f \circ f)(4)$

4.4　VIEWING TRANSFORMATIONS AS FUNCTIONS

KEY IDEAS

Mathematical functions can be used to describe how objects are flipped (reflected), turned (rotated), slid (translated), and changed in size (dilated). These types of functions are called **transformations**. The set of points of the original figure is in the *domain*, and the corresponding set of points of the new figure is in the *range*. Each point of the new figure is the **image** of exactly one point of the original figure, called the **preimage**.

Transformations as Functions

Under a transformation, each point of a figure is "moved" according to some given rule that may be expressed as a function. This function maps the set of points of the original figure onto the set of image points in such a way that each point of the original figure is paired with exactly one point of the new figure. Furthermore, each point of the new figure (range) corresponds to exactly one point of the original figure (domain). Each type of transformation uses a different function rule for "transforming" the original figure into the new figure.

MATH FACTS

DEFINITION OF TRANSFORMATION

A **transformation** is a one-to-one function whose domain and range are the set of points in the plane.

Line Reflections

Reflecting a figure in a line means flipping it over the line so that its mirror image lies across the line and is the same distance from the line. In Figure 4.3, the reflection of $\triangle ABC$ in line k is $\triangle A'B'C'$. This may be written in function notation as

$$r_k(\triangle ABC) = \triangle A'B'C'.$$

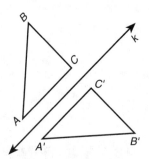

Figure 4.3 Reflecting a Figure in a Line

The function rule associated with this mapping imposes the condition that the image points be located so that line k is the perpendicular bisector of segments $\overline{AA'}, \overline{BB'}$, and $\overline{CC'}$. As a result, each angle and each side of $\triangle ABC$ have the same measure as the corresponding parts of $\triangle A'B'C'$. The two triangles are, therefore, congruent. Thus, under a line reflection, segment length and angle measure are unchanged.

If you point to vertex A of $\triangle ABC$, then to vertex B, and then to vertex C, in that order, your finger is moving in a *clockwise* direction. Doing the same with the corresponding vertices of $\triangle A'B'C'$, your finger moves in a *counterclockwise* (reverse) direction. The *orientations* of the two triangles are reversed. It is this property of reflections that accounts for mirror images of objects appearing "backward" in comparison to the original figure.

MATH FACTS

PROPERTIES OF LINE REFLECTIONS

- Preserve collinearity and betweenness (ordering) of points.
- Preserve distance and angle measure.
- Produce an image congruent to the original figure.
- Do *not* preserve orientation.

Rotations

Rotating a figure means turning it a specified number of degrees about some fixed point, called the **center of rotation**. The amount of turn is called the

angle of rotation. *Positive* angles of rotation turn figures in a *counterclockwise* direction. *Negative* angles of rotation produce *clockwise* turns. In Figure 4.4, the image of $\triangle ABC$ after it is rotated $x°$ in the counterclockwise direction about center O is $\triangle A'B'C'$. Triangle $A'B'C'$ is located in such a way that:

$m\angle AOA' = x$ and $OA = OA'$
$m\angle BOB' = x$ and $OB = OB'$
$m\angle COC' = x$ and $OC = OC'$

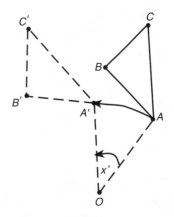

Figure 4.4 Rotating a Figure

In symbols,

$$R_{O,x°}(\triangle ABC) = \triangle A'B'C'.$$

Notice that a rotation is represented by a capital R. The center and the amount of rotation are written as subscripts after the letter R and one-half line below it. Sometimes $R_{O,x°}$ is written as $\mathbf{Rot}_{O,x°}$. When a particular point is understood to be the center of rotation, it may be convenient to omit it and to write $R_{x°}$ instead of $R_{O,x°}$.

Translations

A translation "moves" a figure a fixed distance in a specified direction. A figure and its translated image are congruent. If point (x, y) is translated h units in the horizontal direction and k units in the vertical direction, as shown in Figure 4.5, then its image is determined according to the function rule:

$$T_{h,k}(x,y) = (x + h, y + k).$$

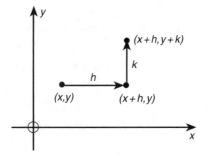

Figure 4.5 Translating a Figure

Examples

1. What is the image of point $(-3, 4)$ under the translation $T_{5,-1}$?

Solution: $T_{5,-1}(-3, 4) = (-3 + 5, 4 + (-1)) = (2, 3)$

2. If the image of $(-4, 1)$ under a translation is $(0, -4)$, what are the coordinates of the image of $(2, 3)$ under the same translation?

Solution: Since $T_{h,k}(-4, 1) = (-4 + h, 1 + k) = (0, -4)$,

$$-4 + h = 0 \quad \text{or} \quad h = 4, \quad \text{and} \quad 1 + k = -4 \quad \text{or} \quad k = -5.$$

Hence $T_{4,-5}(2, 3) = (2+4, 3-5) = \mathbf{(6, -2)}$.

Dilations

A **dilation** of a figure is a size transformation that changes distance while preserving angle measure. The size transformation of a figure that results from viewing it with a magnifying glass is an example of a dilation. Reflections, rotations, and translations produce images *congruent* to the original figures, while dilations produce *similar* figures.

Under a dilation with center O $(0, 0)$ and magnitude k, the image of $A(x, y)$ is $A'(kx,\ ky)$, as can be indicated by writing

$$D_k A(x, y) = A'(kx, ky).$$

The scale factor k can be any nonzero real number. As shown in Figure 4.6, if $k < 0$, then A' is located on the ray opposite $\overrightarrow{OA}$.

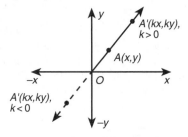

Figure 4.6 Dilating a Figure

MATH FACTS

SIZE TRANSFORMATIONS OF GRAPHS

To find an equation of the dilation of a graph with positive scale factor k, replace x by $\frac{x}{k}$ and y by $\frac{y}{k}$ in the original equation of the graph. For example, the image of the circle $x^2 + y^2 = 9$ under a dilation with a scale factor of 2 is the circle $\left(\frac{x}{2}\right)^2 + \left(\frac{y}{2}\right)^2 = 9$ or, in simpler form, $x^2 + y^2 = 36$.

Examples

3. What are the coordinates of the image of $(-3, 1)$ under dilation D_2?

Solution: In general, $D_2(x, y) = (2x, 2y)$. Hence

$$D_2(-3, 1) = [2 \cdot (-3), 2 \cdot 1] = \mathbf{(-6, 2)}.$$

4. If $D_k(9, 3) = (6, 2)$, what is the constant of dilation?

Solution: In general, $D_k (a, b) = (ka, kb)$. Since $D_k (9, 3) = (9k, 3k) = (6, 2)$ then $9k = 6$ and $3k = 2$. Solve either equation for k:

$$3k = 2$$
$$k = \frac{2}{3}$$

5. What is an equation of the image of the graph of the circle $x^2 + y^2 = 64$ under a size transformation of magnitude $\frac{1}{2}$?

Solution: An equation of the image of the original circle has the general form

$$\left(\frac{x}{k}\right)^2 + \left(\frac{y}{k}\right)^2 = 64.$$

Since the magnitude of the transformation is $\frac{1}{2}$, let $k = \frac{1}{2}$.

$$\left(\frac{x}{\frac{1}{2}}\right)^2 + \left(\frac{y}{\frac{1}{2}}\right)^2 = 64$$
$$(2x)^2 + (2y)^2 = 64$$
$$4x^2 + 4y^2 = 64$$
$$x^2 + y^2 = \frac{64}{4} = 16$$

Hence an equation of the image of the original circle is $x^2 + y^2 = \mathbf{16}$.

Exercise Set 4.4

1. Determine the coordinates of the image of point $(2, -1)$ under each of the following translations:
(a) $T_{0,3}$ (b) $T_{-3,0}$ (c) $T_{-2,4}$ (d) $T_{-5,1}$ (e) $T_{-4,-2}$

2. If the image of $(-1, 5)$ under a translation is $(2, 1)$, find the coordinates of the image of each point under the same translation:
(a) $(-1, 2)$ (b) $(1, -5)$ (c) $(3, 4)$ (d) $(4, 3)$ (e) $(-6, -2)$

3. Find the coordinates of the image of $(-8, -6)$ under each of the following dilations:
(a) D_2 (b) $D_{\frac{1}{2}}$ (c) D_1 (d) D_{-3}

4. If the image of $(4, 6)$ under a certain dilation is $(6, 9)$, find the coordinates of the image of each point under the same dilation:
(a) $(0, 2)$ (b) $(8, 0)$ (c) $(10, 4)$ (d) $(-2, -4)$ (e) $(6, 9)$

4.5 REFLECTING AND ROTATING USING COORDINATES

The coordinates of the images of certain types of reflections and rotations in the coordinate plane can be easily located using special function mapping rules.

Reflecting Using Coordinates

In Figure 4.7, point $A(a, b)$ is reflected in the origin. Its image is determined according to the mapping

$$A(a, b) \rightarrow A'(-a, -b).$$

When this mapping is used, the origin is the midpoint of $\overline{AA'}$.

The mapping rules for reflecting points in special lines in the coordinate plane are summarized in Table 4.2 and illustrated in Figure 4.8.

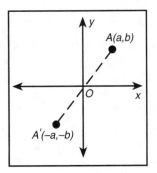

Figure 4.7 Reflecting Using Coordinates

TABLE 4.2 REFLECTING POINTS IN SPECIAL LINES IN THE COORDINATE PLANE

Line of Reflection	Mapping Rule
1. x-axis 2. y-axis	1. $(a, b) \rightarrow (a, -b)$ 2. $(a, b) \rightarrow (-a, b)$
3. $y = k$ 4. $x = h$	3. $(a, b) \rightarrow [a, (2k - b)]$ 4. $(a, b) \rightarrow [(2h - a), b]$
5. $y = x$ 6. $y = -x$	5. $(a, b) \rightarrow (b, a)$ 6. $(a, b) \rightarrow (-b, -a)$

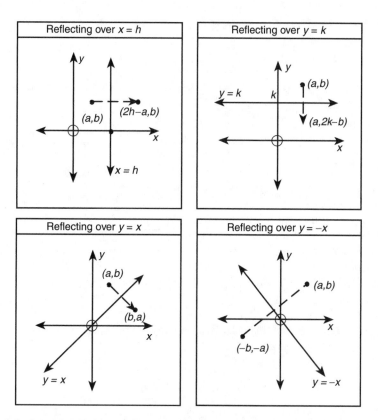

Figure 4.8 Graphing Reflections of (a, b) over the Lines $x = h$, $y = k$, $y = x$, and $y = -x$. If $h = 0$, the line of reflection is the y-axis and the coordinates of the image point simplify to $(-a, b)$. If $k = 0$, the line of reflection is the x-axis and the coordinates of the image point simplify to $(a, -b)$.

Example

1. Find the image of a reflection of $A(3, 4)$ in each of the following lines:
(a) $y = -x$ **(b)** $x = 2$ **(c)** $y = -1$

Solutions: **(a)** If (a, b) is reflected in the line $y = -x$, its image is $(-b, -a)$. Let $a = 3$ and $b = 4$. Then

$$(3, 4) \xrightarrow{\;r_{y=-x}\;} (-4, -3).$$

(b) If (a, b) is reflected in the vertical line $x = h$, its image is $[(2h - a), b]$. Let $h = 2$, $a = 3$, and $b = 4$. Then

$$(3, 4) \xrightarrow{\;r_{x=2}\;} [(2 \cdot 2 - 3), 4)] = (1, 4).$$

(c) If (a, b) is reflected in the horizontal line $y = k$, its image is $[(a, (2k - b)]$. Let $k = -1$, $a = 3$, and $b = 4$. Then

$$(3, 4) \xrightarrow{\;r_{y=-1}\;} [3, (2(-1) - 4)] = (3, -6).$$

Rotating Using Coordinates

In Figure 4.9, rectangle $AB'C'D'$ is the image of rectangle $ABCD$ under a 90° counterclockwise rotation about the origin. Compare the coordinates of corresponding vertices of the two rectangles. For example,

$$C(6, 3) \rightarrow C'(-3, 6).$$

In general, under a 90° rotation (a, b) is mapped onto $(-b, a)$.

The coordinates of the images of points rotated about the origin through angles that are multiples of 90° can be determined using the function rules given in Table 4.3.

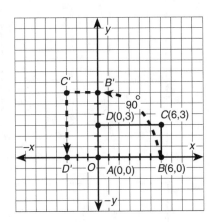

Figure 4.9 Rotation Through Angles That Are Multiples of 90°

TABLE 4.3 FUNCTION RULES FOR SPECIAL ROTATIONS

$$R_{90°}(a, b) = (-b, a)$$
$$R_{180°}(a, b) = (-a, -b)$$
$$R_{270°}(a, b) = (b, -a)$$

You should verify the following:

- Rotating a point 360° about the origin produces an image that coincides with the original point.
- Rotating a point –90° about the origin is equivalent to rotating the point 270° about the origin. In general, a rotation of x degrees in the clockwise direction ($x < 0$) is equivalent to a rotation of $360 - |x|$ degrees in the counterclockwise direction.

Examples

2. What is the image of point (–1, 5) under a rotation of 90° about the origin in the counterclockwise direction?

Solution: In general, $R_{90°}(x, y) = (-y, x)$. Hence **(–1, 5) → (–5, –1)**.

3. Under a counterclockwise rotation about the origin, the image of (–2, 4) is (4, 2). What is the *smallest* possible measure of the angle of rotation?

Solution: The coordinates of the image of (–2, 4) are determined according to the rule $(x, y) \rightarrow (y, -x)$, where $x = -2$ and $y = 4$. This mapping rule defines a counterclockwise rotation of **270°** about the origin.

4. Which rotation is equivalent to $R_{-150°}$?

(1) $R_{150°}$ (2) $R_{-210°}$ (3) $R_{210°}$ (4) $R_{510°}$

Solution: A rotation of $-150°$ is equivalent to a rotation of $360 - |-150|$ $= 360 - 150 = 210°$. The correct choice is **(3)**.

Half-Turns

Rotating point $A(a, b)$ 180° about the origin and reflecting point A in the origin produce images having the same coordinates, namely, $(-a, -b)$. These two transformations are, therefore, equivalent. A rotation of 180° about any point P is called a **half-turn** and is equivalent to a reflection in point P.

In Figure 4.10, the image of a half-turn of $\overline{AB}$ about center P is the line segment whose endpoints are the reflections of points A and B in point P. Points A' and B' are determined so that P is the midpoint of $\overline{AA'}$ and $\overline{BB'}$.

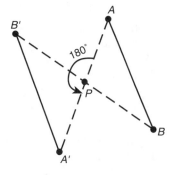

Figure 4.10 A Half-Turn
$R_{P,180°}:\overline{AB} \rightarrow \overline{A'B'}$

Exercise Set 4.5

1. Find the image of a reflection of $P(-1, 3)$ in each of the following:

(a) $(0, 0)$ (c) y-axis   (e) $y = -1$
(b) x-axis   (d) $y = x$ (f) $x = 2$

2. Find the image of a reflection of $A(1, 4)$ in each line:

(a) $x = 0$ (c) $y = x$ (e) $y = 2$
(b) $y = 0$ (d) $y = -x$ (f) $x = -1$

3. Find the image of each point after a reflection in the line $x = 3$ followed by a reflection in the origin:

(a) $(1, 1)$ (b) $(-1, 2)$ (c) $(3, -1)$ (d) $(2, 0)$ (e) $(0, 0)$

4. Find the image of each point after a reflection in the line $y = 1$ followed by a reflection in the line $y = x$:

(a) $(3, 3)$ (b) $(2, -1)$ (c) $(-1, 4)$ (d) $(0, 1)$ (e) $(0, 0)$

5. For each rotation, find the *smallest* positive angle of rotation that produces an equivalent rotation about the same center:

(a) $R_{-180°}$ (b) $R_{-230°}$ (c) $R_{-65°}$

6. Find the image of point $(3, -2)$ under each rotation about the origin:
   (a) $R_{90°}$ (b) $R_{180°}$ (c) $R_{270°}$ (d) $R_{360°}$

7. Find the image of point $(-4, 1)$ under each rotation about the origin:
   (a) $R_{-90°}$ (b) $R_{-180°}$ (c) $R_{-270°}$ (d) $R_{-360°}$

8. Which transformation does *not* preserve distance?
   (1) dilation   (2) translation   (3) rotation   (4) reflection

9. Which transformation of $P(x, y)$ is *not* equivalent to the other three transformations?
   (1) a reflection of $P(x, y)$ in the origin
   (2) the translation $P(x, y) \rightarrow P'(x - 1, y - 1)$
   (3) a dilation of $P(x, y)$ with center at the origin and a scale factor of -1
   (4) a rotation of $P(x, y)$ $180°$ about the origin

10. What is an equation of the line that passes through the origin and contains the image of $T_{2,3}(1, 3)$?

11. A sequence of transformations that reflect point P through the origin, over the x-axis, and then over the line $y = x$ is equivalent to a rotation about the origin. What is the amount of the rotation?

12. (a) On graph paper, draw and label the line segment determined by $A(4, 1)$ and $B(5, 4)$.
    (b) Graph $\overline{A'B'}$, the image of $\overline{AB}$ after a reflection over line $y = x$.
    (c) Graph $\overline{A''B''}$, the image of $\overline{A'B'}$ after the transformation defined by the mapping $(x, y) \rightarrow (x - 5, y - 5)$.
    (d) Graph $\overline{A'''B'''}$, the image of $\overline{A''B''}$ after a reflection through the origin.
    (e) Write a translation that will map $\overline{A'B'}$ onto $\overline{B'''A'''}$.

13. (a) On graph paper, draw and label $\triangle ABC$, whose vertices are $A(3, 1)$, $B(4, 3)$, and $C(6, 1)$.
    (b) Draw the image of $\triangle ABC$ under a reflection in the line $y = x$. What are the coordinates of $\triangle A'B'C'$, the image of $\triangle ABC$?
    (c) Draw the image of $\triangle A'B'C'$ under a rotation about the origin of $90°$ in the counterclockwise direction. What are the coordinates of $\triangle A''B''C''$, the image of $\triangle A'B'C'$?
    (d) Draw the image $\triangle A''B''C''$ under a reflection in the origin. What are the coordinates of $\triangle A'''B'''C'''$, the image of $\triangle A''B''C''$?

4.6 COMPOSING TRANSFORMATIONS

Geometric transformations, like algebraic functions, may be composed. A composite transformation is evaluated by working from right to left. For example, to evaluate $r_{x\text{-axis}} \circ r_{y\text{-axis}}$ begin by reflecting in the y-axis:

$$(x, y) \xrightarrow{\;r_{y\text{-axis}}\;} (-x, y) \xrightarrow{\;r_{y\text{-axis}}\;} (-x, -y).$$

Some composite transformations produce familiar transformations. For example, the image point $(-x, -y)$ is also obtained by reflecting (x, y) in the origin. Hence the composite of reflections in the coordinate axes is equivalent to a reflection in the origin.

Evaluating Composite Transformations

In composing transformations, the order in which the transformations are performed matters, as Example 1 illustrates. Example 2 shows that more than two functions may be composed.

Examples

1. Find the coordinates of the image of $P(-2, 4)$ under each composite transformation:

(a) $r_{y = x} \circ r_{y\text{-axis}}$ **(b)** $r_{y\text{-axis}} \circ r_{y = x}$

Solutions: **(a)** In general,

$$r_{y\text{-axis}}(x, y) = (-x, y) \quad \text{and} \quad r_{y = x}(x, y) = (y, x).$$

Perform $r_{y\text{-axis}}$ first, followed by $r_{y = x}$:

$$(-2, 4) \xrightarrow{\;r_{y\text{-axis}}\;} (2, 4) \xrightarrow{\;r_{y = x}\;} (4, 2).$$

Under the composite transformation, the coordinates of the image of $P(-2, 4)$ are **(4, 2)**.

(b) Perform $r_{y = x}$, followed by $r_{y\text{-axis}}$:

$$(-2, 4) \xrightarrow{\;r_{y = x}\;} (4, -2) \xrightarrow{\;r_{y\text{-axis}}\;} (-4, -2).$$

Under the composite transformation, the coordinates of the image of $P(-2, 4)$ are **(−4, −2)**.

2. In the accompanying figure, p and q are lines of symmetry for regular hexagon $ABCDEF$ intersecting at point O, the center of the hexagon. Determine the image of each composite transformation:

(a) $r_p \circ r_q (\overline{AB})$ (b) $r_q \circ r_p \circ r_q (D)$

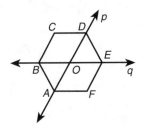

Solutions: **(a)** Reflect $\overline{AB}$ in line q, followed by the reflection of its image in line p.

$$\overline{AB} \xrightarrow{r_q} \overline{CB} \xrightarrow{r_p} \overline{EF}$$

(b) When the composite is evaluated from right to left, point D is reflected in line q, followed by a reflection in line p, followed by a reflection in line q.

$$D \xrightarrow{r_q} F \xrightarrow{r_p} B \xrightarrow{r_q} B$$

3. If $(9, -1)$ is the image under the composite transformation $T_{h,k} \circ T_{4,3}$ $(2, -2)$, find the values of h and k.

Solution:

$$(2, -2) \xrightarrow{T_{4,3}} (6, 1) \xrightarrow{T_{h,k}} (6 + h, 1 + k) = (9, -1)$$

Hence $6 + h = 9$ and $1 + k = -1$, so $h = 3$ and $k = -2$.

Glide Reflections

Sliding a reflected image in a direction parallel to the line of reflection produces a *glide reflection*, as shown in Figure 4.11.

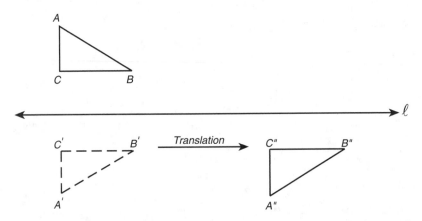

Figure 4.11 Glide Reflection Triangle $A''B''C''$ is the image of $\triangle ABC$ after a reflection in line ℓ followed by a translation in a direction parallel to line ℓ. Triangle $A''B''C''$ is also the image of $\triangle ABC$ when the reflection follows the translation.

DEFINITION OF GLIDE REFLECTION

A **glide reflection** is the composite of a line reflection and of a translation whose direction is parallel to the reflecting line.

Since any translation may be expressed as the composition of reflections in parallel lines, a glide reflection may be thought of as the composition of reflections in parallel lines and a third reflection in a line that is perpendicular to the parallel lines.

Isometry

A line reflection, a rotation, and a translation each preserves distance and is, therefore, an *isometry* (plural: *isometries*). A dilation, however, is *not* an isometry since it does *not* preserve distance.

DEFINITION OF ISOMETRY

An **isometry** is a transformation that preserves distance.

If an isometry preserves orientation, it is called a **direct isometry**. Rotations and translations preserve orientation and are, therefore, direct isometries. An **opposite isometry** is an isometry in which orientation changes. A line reflection reverses orientation and is, therefore, an opposite isometry.

Examples

4. Which transformation is *not* an isometry?

(1) $(x, y) \rightarrow (x - 1, y + 2)$ (3) $(x, y) \rightarrow \left(\dfrac{1}{2}x, \dfrac{1}{2}y \right)$

(2) $(x, y) \rightarrow (x, -y)$ (4) $(x, y) \rightarrow (-y, -x)$

Solution: An isometry preserves distance. Choice (1) is a translation, choice (2) is a reflection in the *x*-axis, choice (4) is a reflection in the line $y = -x$. Each of these transformations preserves distance.

The transformation described by the mapping $(x, y) \rightarrow \left(\dfrac{1}{2}x, \dfrac{1}{2}y\right)$ is a dilation, which does *not* preserve distance since the scale factor is $\dfrac{1}{2}$. The correct choice is **(3)**.

Exercise Set 4.6

1. Find the coordinates of the image of each point under the composite transformation $r_{y = x} \circ r_{x\text{-axis}}$:
   (a) $(4, 7)$ (b) $(5, -2)$ (c) $(0, 3)$ (d) $(-4, 0)$ (e) $(-2, -3)$

2. Find the coordinates of the image of each point under the composite transformation $r_{y = 1} \circ r_{y\text{-axis}}$:
   (a) $(2, 0)$ (b) $(0, -1)$ (c) $(5, -1)$ (d) $(-2, 5)$ (e) $(-3, -4)$

3. Given point $A(-2, 5)$, find the coordinates of the image of each composite transformation:
   (a) $T_{2,2} \circ T_{-1,3}(A)$
   (b) $T_{-1,0} \circ T_{0,-1}(A)$
   (c) $R_{90°} \circ r_{y = x}(A)$
   (d) $r_{x\text{-axis}} \circ T_{2,-4}(A)$
   (e) $T_{2,-3} \circ r_{y = 1}(A)$
   (f) $r_{x\text{-axis}} \circ r_{y = x} \circ r_{y\text{-axis}}(A)$

4. If $(-4, 8)$ is the image under the composite transformation $T_{h,3} \circ T_{-2,k}(-3, 0)$, find the values of h and k.

5. Which transformation is *not* an isometry?
   (1) $J: (x, y) \rightarrow (x + 2, y + 2)$
   (2) $K: (x, y) \rightarrow (2x, 2y)$
   (3) $L: (x, y) \rightarrow (y, -x)$
   (4) $M: (x, y) \rightarrow (-x, -y)$

6. Which transformation is an opposite isometry?
   (1) $S: (x, y) \rightarrow (x - 2, y - 2)$
   (2) $T: (x, y) \rightarrow (y, x)$
   (3) $U: (x, y) \rightarrow (-y, x)$
   (4) $W: (x, y) \rightarrow (y, -x)$

7. Which composite transformation does *not* represent a glide reflection?
   (1) $r_{x\text{-axis}} \circ T_{4,0}$
   (2) $r_{y\text{-axis}} \circ T_{0,4}$
   (3) $r_{y = x} \circ T_{2,4}$
   (4) $r_{x = 1} \circ r_{y = 3} \circ r_{x = 5}$

8. In the accompanying figure, lines ℓ and m are lines of symmetry. Find:
  (a) $r_m \circ r_\ell(FG)$ **(b)** $r_\ell \circ r_m(BC)$ **(c)** $r_m \circ r_\ell \circ r_m(H)$

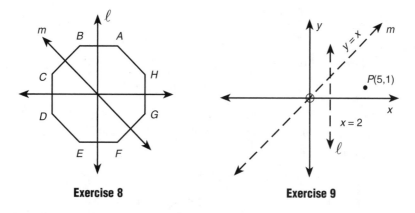

Exercise 8   Exercise 9

9. In the accompanying figure, lines ℓ and m are lines of symmetry
where line m refers to the line $y = x$. Find:
  (a) $r_\ell \circ r_m(P)$ **(b)** $r_m \circ r_\ell(P)$ **(c)** $r_\ell \circ r_m \circ r_{x\text{-axis}}(P)$

10. (a) On graph paper, graph and label the triangle whose vertices are
       $A(6, 1)$, $B(8, 5)$, and $C(5, 4)$.
   (b) Graph and state the coordinates of $\triangle A'B'C'$, the image of $\triangle ABC$
       after a reflection over the line $y = x$.
   (c) Graph and state the coordinates of $\triangle A''B''C''$, the image of $\triangle ABC$
       after the composite transformation $r_{y\text{-axis}} \circ r_{x\text{-axis}} (\triangle ABC)$.
   (d) Find the coordinates of $\triangle A'''B'''C'''$, the image of $\triangle ABC$ after a
       translation that maps $P(0,0)$ onto $P'(3, -1)$.

11. (a) On graph paper, graph and label the triangle whose vertices are
       $A(4, 0)$, $B(8, 1)$ and $C(8, 4)$.
   (b) Graph and state the coordinates of $\triangle A'B'C'$, the image of $\triangle ABC$
       after the composite transformation $r_{x=0} \circ r_{y=x}(\triangle ABC)$.
   (c) Which single type of transformation maps $\triangle ABC$ onto $\triangle A'B'C'$?
       (1) rotation   (3) glide reflection
       (2) dilation   (4) translation
   (d) Graph and state the coordinates of $\triangle A''B''C''$, the image of $\triangle ABC$
       after the composite transformation $r_{y=-4} \circ r_{y=0}(\triangle ABC)$.
   (e) Which single type of transformation maps $\triangle ABC$ onto $\triangle A''B''C''$?
       (1) rotation   (3) glide reflection
       (2) dilation   (4) translation

4.7 FINDING THE INVERSE OF A FUNCTION

$$\bigwedge \text{KEY IDEAS} \bigwedge$$

In each ordered pair of the function

$$f = \{(1,2),(2,4),(3,6),(4,8)\}.$$

the y-value is two times the x-value. To undo or reverse the effect of this function, start with the y-value, and halve it to get back the original x-value. The result is the function

$$f^{-1} = \{(2,1), (4,2), (6,3),(8,4)\}.$$

The -1 in f^{-1} is *not* an exponent. The expression f^{-1} is read as "the **inverse** of function f." The inverse of a function is formed by interchanging the x- and y-members of each ordered pair of the function.

Forming Inverse Functions

To form the inverse of a function, switch the roles of x and y.

- If a function is described by a set of ordered pairs, form its inverse by switching the positions of the x- and y-values in each ordered pair. Thus, if f = {(−1, 0), (0, 1), (1, 2)}, then the inverse of function f is the function

$$f^{-1} = \{(0, -1), (1, 0), (2, 1)\}.$$

- If a function is described by an equation, form its inverse by (1) interchanging x and y; and then (2) solving for y in terms of x. To find the inverse of $2x - y = 3$, proceed as follows:

Interchange x and y: $2y - x = 3$
Solve for y: $2y = x + 3$
$$y = \frac{x+3}{2}$$

Example

1. Find the inverse of $f(x) = x - 2$.

Solution:
Replace $f(x)$ by y: $y = x - 2$

Interchange x and y: $x = y - 2$
Solve for y: $y = x + 2$
Replace y by $f^{-1}(x)$: $f^{-1}(x) = x + 2$ or $y = x + 2$

One-to-One Functions

If $f = \{(2, 4), (3, 9)\}$, then its inverse is $f^{-1} = \{(4, 2), (9, 3)\}$. Since function f has the property that no two of its ordered pairs have the same y-value, its inverse f^{-1} is a function. The inverse of a function, however, is not always a function. For example, if $g = \{(1, 5), (3, 5)\}$, then its inverse is $\{(5, 1), (5, 3)\}$. Although g is a function, its inverse is not a function.

Functions that have inverses that are functions are called *one-to-one functions*. Function f is a one-to-one function, but function g is *not* a one-to-one function.

MATH FACTS

DEFINITION OF ONE-TO-ONE FUNCTION

A **one-to-one function** is a function in which no two ordered pairs have the same y-value.

It follows that:

- If a function is one-to-one, then its inverse is also a function.
- If a function is *not* one-to-one, then its inverse is *not* a function.

Example

2. Determine whether each function is a one-to-one function. If the function is one-to-one, determine its inverse.
(a) $y = 4x$ **(b)** $y = |x|$

Solutions: **(a)** The function described by the equation $y = 4x$ is one-to-one, since no value of x that is substituted into the equation $y = 4x$ ever produces two different values for y. The inverse is formed by interchanging x and y and then solving for y:

$$x = 4y$$
$$y = \frac{x}{4}$$

(b) The absolute-value function includes ordered pairs such as (2, 2) and (−2, 2). Hence $y = |x|$ does *not* describe a one-to-one function, so its inverse is *not* a function.

Horizontal-Line Test

Since, in Figure 4.12, a horizontal line intersects the accompanying parabola in two different points, the graph contains two ordered pairs with the same y-value. Thus, the graph does *not* represent a one-to-one function.

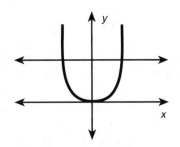

Figure 4.12 Horizontal-Line Test

MATH FACTS

HORIZONTAL-LINE TEST

A function is a one-to-one function if no horizontal line intersects its graph in more than one point. The graph at the right passes the horizontal-line test and therefore has an inverse function.

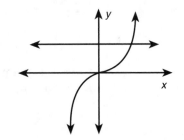

Symmetry and Inverse Functions

For every point (a, b) contained on the graph of a one-to-one function, the corresponding point (b, a) is found on the graph of its inverse, as shown in Figure 4.13. Also, the reflection of point (a, b) in the line $y = x$ is point (b, a). Thus, the graphs of a function and its inverse are symmetric with respect to the line $y = x$, and either graph may be obtained by reflecting the other graph in the line $y = x$.

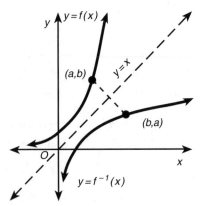

Figure 4.13 Symmetry and Inverse Functions

92

Example

3. If $f(x) = x^2$ and $x \geq 0$, find $f^{-1}(x)$.

Solution: Since the domain is restricted to nonnegative values, the graph of $f(x) = x^2$ is confined to Quadrant I, as shown in the acompanying diagram. The graph passes the horizontal-line test, so proceed to find the inverse function.

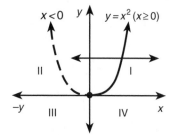

Write original equation: $y = x^2$
Interchange x and y: $x = y^2$
Solve for y: $y = \pm\sqrt{x}$

Since the domain of the original function (nonnegative numbers) becomes the range of the inverse function, y must also be nonnegative. Hence drop the $\pm$ sign in front of $\sqrt{x}$ and write $y = \mathbf{f^{-1}(x)} = \sqrt{x}$.

Composing Inverse Functions

Composing a function $f(x)$ and its inverse, in any order, produces the starting value, x. Thus functions f and g are inverses if $f(g(x)) = x$ and $g(f(x)) = x$.

Examples

4. Prove that $f(x) = 1 - x^3$ and $g(x) = \sqrt[3]{1-x}$ are inverse functions.

Solution: Show that $f(g(x)) = x$ and $g(f(x)) = x$.

$$f(g(x)) = 1 - [g(x)]^3 \qquad\qquad g(f(x)) = \sqrt[3]{1 - f(x)}$$
$$= 1 - (\sqrt[3]{1-x})^3 \qquad\qquad\quad = \sqrt[3]{1 - (1 - x^3)}$$
$$= 1 - (1 - x) \qquad\qquad\qquad\quad = \sqrt[3]{x^3}$$
$$= x \qquad\qquad\qquad\qquad\qquad\quad = x$$

Since $f(g(x)) = g(f(x)) = x$, **f and g are inverses**.

5. If $f(x) = 1.9x^3 - 17$, find $f(f^{-1}(6.4))$.

Solution: Since $f(f^{-1}(x)) = x$, then $f(f^{-1}(6.4)) = \mathbf{6.4}$.

Exercise Set 4.7

1–8. For each function, find its inverse, provided that it exists. If it does not exist, tell why.

1. $f = \{(-1, 2), (2, 3), (4, 5), (-2, 3)\}$

2. $g = \{(-2, 5), (5, -2), (1, 2)\}$

3. $y = 3 - x$

4. $g(x) = x^3 + 1$

5. $f(x) = 3x + 2$

6. $h(x) = \sqrt{x^2 + 1}$

7. $2x + 3y = 5$

8. $f(x) = \dfrac{x^2}{2} + 4$

9 and 10. For each pair, prove that the functions are inverse.

9. $f(x) = 7x + 1$ and $g(x) = \dfrac{x - 1}{7}$

10. $f(x) = 2 - \dfrac{1}{x}$ and $g(x) = \dfrac{1}{x - 2}$

11. Which transformation can be used to find the set of ordered pairs of the inverse of a one-to-one function?

12–14. Determine whether the function represented by each graph has an inverse function.

12.   **13.**   **14.**

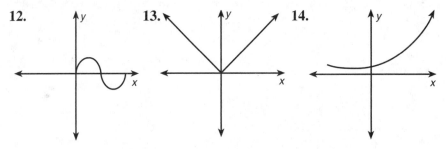

4.8 GRAPHING AN INVERSE VARIATION

KEY IDEAS

If $xy = 48$, then, as x *increases* from 1 to 2 to 3 to 4 . . . , the corresponding value of y *decreases* from 48 to 24 to 16 to 12 . . . . Variables x and y are said to be *inversely* related since, as the value of one variable gets larger, the value of the other variable becomes smaller. This happens whenever the product of x and y remains constant.

Definition of Inverse Variation

An **inverse variation** between two nonzero variables, say x and y, is a relation in which their product remains constant. If k is some nonzero constant, then y varies inversely as x if

$$xy = k \quad \text{or, equivalently,} \quad y = \frac{1}{4} \quad (k \neq 0).$$

In these equations, k is called the **constant of variation.**

Example

1. If y varies inversely as x, and $y = 32$ when $x = 6$, find y when $x = 9$.

Solution: If y varies inversely as x, then the product of x and y must be the same for any two ordered pairs (x_1, y_1) and (x_2, y_2). Thus

$$x_1 \cdot y_1 = x_2 \cdot y_2$$

Let $x_1 = 6$, $y_1 = 32$, and $x_2 = 9$: $\quad (6)(32) = (9)(y)$

Simplify: $\qquad\qquad\qquad\qquad\qquad 198 = 9y$

Divide both sides by 9: $\qquad\qquad \dfrac{198}{9} = \dfrac{9y}{9}$

$$22 = y \quad \text{or} \quad y = \mathbf{22}$$

Graph of an Inverse Variation

The graph of $xy = 4$, shown in Figure 4.14, is called an **equilateral** or **rectangular hyperbola**. The hyperbola consists of two different branches located in opposite quadrants. Since the graph passes the vertical-line test, the equation $xy = 4$ describes a function.

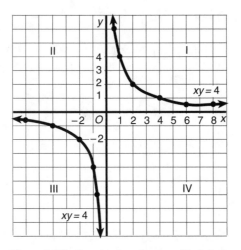

Figure 4.14 Graph of an Inverse Variation

In general, the graph of $xy = k$ has these additional features:

- The coordinate axes are *asymptotes*—lines that frame the graph and to which the graph gets closer and closer but never touches.
- If $k > 0$, the branches of the graph are located in Quadrants I and III, since in these quadrants the x- and y-coordinates have the same sign, so their product is positive.
- If $k < 0$, the branches of the graph are located in Quadrants II and IV, since in these quadrants the x- and y-coordinates have opposite signs, so their product is negative.
- Each branch of the hyperbola is a reflection of the other branch in the origin. If point (a, b) lies on a branch of the hyperbola, then its image on the other branch is point $(-a, -b)$.

Examples

2. (a) Draw the graph of $xy = 6$ in the interval $-6 \le x \le 6$ $(x \ne 0)$.

   (b) On the same set of axes, sketch the graph of the image of $xy = 6$ after a rotation of 90°.

Solutions: **(a)** On the closed interval $-6 \le x \le 6$ $(x \ne 0)$, the graph of $xy = 6$ is in Quadrants I and III and has the y-axis as an asymptote, $x = 6$ as an endpoint in Quadrant I, and $x = -6$ as an endpoint in Quadrant III. Construct a table of ordered pairs, using all integer values of x between -6 and 6, inclusive (except $x \ne 0$). Find the corresponding value of y such that $xy = 6$.

	Quadrant I				Quadrant III			
x	1	2	3	6	−1	−2	−3	−6
y	6	3	2	1	−6	−3	−2	−1

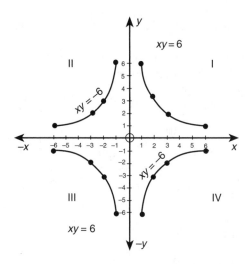

(b) Under a counterclockwise rotation of 90° about the origin, each point (x, y) of the graph is mapped onto $(-y, x)$. Replacing x by $-y$ and y by x in $xy = 6$ gives $(-y)(x) = 6$ or, equivalently, $xy = -6$. The graph of $xy = -6$ is located in Quadrants II and IV and has the restricted range $-6 \leq y \leq 6$ $(y \neq 0)$.

You can easily verify this fact by actually rotating the page that contains the graph of $xy = 6$ counterclockwise 90°. When you do this, the image of the x-axis is vertical, and the image of the positive y-axis lies horizontally at the left side of the page.

3. In Example 2, what is the equation of the image of the graph of $xy = 6$ after it is dilated using a scale factor of 2?

Solution: Under a dilation having a scale factor of 2, the image of each point (x, y) of the graph is mapped onto $(2x, 2y)$. The coordinates of each point on the original graph are one-half the coordinates of the corresponding image point. Thus, replacing x by $\frac{x}{2}$ and y by $\frac{y}{2}$ in the original equation gives the equation of the dilated graph:

$$xy = 6 \xrightarrow{D_2} \left(\frac{x}{2}\right)\left(\frac{y}{2}\right) = 6 \text{ or } xy = 24$$

Hence the image of $xy = 6$ is the graph whose equation is **$xy = 24$.** If the domain of the original graph is $-6 \leq x \leq 6$, then the domain of the dilated graph is $-12 \leq x \leq 12$ $(x \neq 0)$, since each value of x is doubled.

Exercise Set 4.8

1. If y varies inversely as x, and $y = 24$ when $x = 3$, find x when $y = 8$.

2. If y varies inversely as x, and $y = 80$ when $x = 5$, find x when $y = 100$.

3. If y varies inversely as x, and $x = 4$ when $y = 21$, find y when $x = 6$.

4. The number of hours required to complete a certain job varies inversely as the number of people assigned to the job. It is known that 8 people take 24 hours to complete the job. How many hours will 12 people need to complete the same job, assuming that each person does the same amount of work?

5. (a) Using graph paper, draw the graph of $xy = 12$, where $-12 \le x \le 12$ $(x \ne 0)$.
   (b) What is an equation of the image of $xy = 12$ under a dilation having a scale factor of $\frac{1}{2}$?
   (c) Sketch the image of $xy = 12$ after a reflection in the x-axis. What is an equation of this graph?

6. (a) Using graph paper, draw the graph of $xy = -10$ in the interval $-10 \le x \le 10$ $(x \ne 0)$.
   (b) On the same set of axes, draw and label the graph of the image of $xy = -10$ after a rotation of $90°$. What is an equation of this graph?

7. The price per person to rent a limousine for a prom varies inversely as the number of passengers. If five people rent the limousine, the cost is $70 each. How many people are renting the limousine when the cost *per couple* is $87.50?

Each of the questions in this section has appeared on a previous Course III Regents Examination. Here is an opportunity for you to review Chapter 4 and, at the same time, prepare for the Course III Regents Examination.

1. What is the image of $R_{-180°}(-4, -7)$?

2. If f$(x) = 3x$ and g$(x) = 7x - 1$, what is (f ∘ g)(4)?

3. If $M(-2, 8)$ is reflected in the y-axis, what are the coordinates of M', the image of M?

4. If f$(x) = x^{\frac{3}{4}}$, find f(16).

5. If f$(x) = 4x^{-2} - 2x^0$, find the value of f(2).

6. A translation maps $A(-2, 1)$ onto $A'(2, 2)$. Find the coordinates of B', the image of $B(-4, -5)$ under the same translation.

7. In the accompanying diagram of a regular octagon, ℓ, m, and p are lines of symmetry. What is $r_p \circ r_m(E)$?

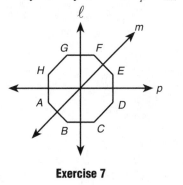

Exercise 7

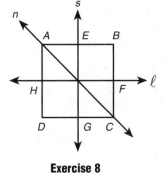

Exercise 8

8. In the accompanying diagram, $ABCD$ is a square with symmetry lines n, s, and ℓ. What is $r_n \circ r_s(\overline{BF})$?

9. In the accompanying figure, p and q are symmetry lines for the figure $ABCDEF$. Find $r_q \circ r_p \circ r_q(A)$.

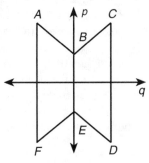

Exercise 9

10. The rate at which a man travels from city A to city B varies inversely as the time it takes to make the trip. If the man can make the trip in $3\frac{1}{2}$ hours at 60 kilometers per hour, how many kilometers per hour must he travel to make the trip in 3 hours?

11. Find the image of $A(4, -3)$ under the transformation $r_{x = 2}$.

12. The diameter of a wheel varies inversely as the number of revolutions that the wheel makes to cover a certain distance. If a wheel with a 26-inch diameter makes 10 revolutions in covering a certain distance, how many revolutions will a wheel with a diameter of 20 inches make in covering the same distance?

13. For any point (x, y), which transformation is equivalent to $R_{45°} \circ R_{-135°}$?
(1) $R_{-90°}$ (3) $r_{y = x}$
(2) $R_{90°}$ (4) $r_{x\text{-axis}}$

14. Which transformation is *not* an isometry?
(1) $(x, y) \rightarrow (x + 6, y - 2)$ (3) $(x, y) \rightarrow \left(\dfrac{x}{2}, \dfrac{y}{2}\right)$
(2) $(x, y) \rightarrow (y, -x)$ (4) $(x, y) \rightarrow (-y, -x)$

15. The domain for $g(x) = 5x - 1$ is $-2 \leq x \leq 2$. The *smallest* value in the range of $g(x)$ is
(1) -11 (2) 9   (3) 11   (4) -9

16. The graph of the equation $4x^2 + 16y^2 = 25$ is
(1) a circle   (3) an ellipse
(2) a hyperbola   (4) a parabola

17. If $f(x) = x^2 - 3$, then $f(a - b)$ is equivalent to:
(1) $a^2 - b^2 - 3$ (3) $a^2 - 2ab + b^2 - 3$
(2) $a^2 - 2ab - b^2 - 3$ (4) $a^2 + b^2 - 3$

18. Given: points $A(3, 0)$ and $B(-4, 6)$. Write the letters (a) through (e), and next to *each* letter write the coordinates of the images of points A and B after each transformation described.
(a) the images of points A and B after a reflection in the line $y = x$
(b) the images of points A and B after a rotation of $90°$ counterclockwise about the origin
(c) the images of points A and B after a reflection in the line $x = 2$
(d) the images of points A and B after a reflection through the origin
(e) the images of points A and B after a dilation $D_{\frac{1}{2}}$

19. Triangle ABC has coordinates $A(-1, 4)$, $B(3, 7)$, and $C(5, 1)$.
   (a) On graph paper, draw and label $\triangle ABC$.
   (b) Graph and state the coordinates of $\triangle A'B'C'$, the image of $\triangle ABC$ under the composition $R_{0,\,90} \circ r_{x\text{-axis}}$.
   (c) State the single transformation equivalent to $R_{0,\,90} \circ r_{x\text{-axis}}$.

ANSWERS TO SELECTED EXERCISES: CHAPTER 4

Section 4.1
   1. Domain = $\{-2, 0, 3\}$, range = $\{2, 3\}$
   3. Domain = range = {all real numbers}
   5. Domain = $\{x: x \geq 3\}$, range = $\{y: y \geq 0\}$
   7. Domain = $\{x: x \geq 0\}$, range = $\{y: y \geq 3\}$
   9. (2)   **10.** (1)
   11. (a) 1   **(c)** 19   **(e)** 11
   12. (a) 12   **(b)** 18   **(c)** 2
   13. -3 and 2
   14. (a) 0   **(b)** 16   **(c)** 18   **(d)** $4y^2 + 4y + 1$ **(e)** x^2

Section 4.2
   1. Circle is not a function.
   2. Hyperbola is not a function.
   3. Circle is not a function.
   4. Parabola is a function.
   5. Line is a function.
   6. Ellipse is not a function.
   7. Circle is not a function.
   8. Parabola is a function.
   9. Hyperbola is not a function.
   10. Graph represents a function.
   11. Graph represents a function.
   12. Graph does not represent a function.
   13. Graph represents a function.

Section 4.3
   1. (a) $\dfrac{x^2}{4}$ **(b)** $\dfrac{x^2}{2}$ **(c)** x^4 **(d)** $\dfrac{x}{4}$
   3. (a) x **(b)** x **(c)** $x^{\frac{1}{4}}$ **(d)** x^4
   5. (a) $x^2 - 2x + 1$ **(b)** $1 - x^2$ **(c)** x^4 **(d)** x
   7. (a) 512   **(b)** $\dfrac{x^9}{512}$ **(c)** a^{18} **9.** 5
   10. (a) -1 **(b)** 3   **(c)** 3   **(d)** Does not exist.   **(e)** -1 **(f)** -1

Section 4.4
   1. (a) $(2, 2)$ **(b)** $(-1, -1)$ **(c)** $(0, 3)$ **(d)** $(-3, 0)$ **(e)** $(-2, -3)$
   2. (a) $(2, -2)$ **(b)** $(4, -9)$ **(c)** $(6, 0)$ **(d)** $(7, -1)$ **(e)** $(-3, -6)$
   3. (a) $(-16, -12)$ **(b)** $(-4, -3)$ **(c)** $(-8, -6)$ **(d)** $(24, 18)$
   4. (a) $(0, 3)$ **(b)** $(12, 0)$ **(c)** $(15, 6)$ **(d)** $(-3, -6)$ **(e)** $\left(9, \dfrac{27}{2}\right)$

Section 4.5
1. **(a)** $(1, -3)$ **(b)** $(-1, -3)$ **(c)** $(1, 3)$ **(d)** $(3, -1)$ **(e)** $(-1, -5)$ **(f)** $(5, 3)$
3. **(a)** $(-5, -1)$ **(b)** $(-7, -2)$ **(c)** $(-3, -1)$ **(d)** $(-4, 0)$ **(e)** $(-6, 0)$
5. **(a)** $180°$ **(b)** $130°$ **(c)** $295°$
7. **(a)** $(1, 4)$ **(b)** $(4, -1)$ **(c)** $(-1, -4)$ **(d)** $(-4, 1)$
9. (2)
11. $270°$
12. **(b)** Coordinates of endpoints are: $A'(1, 4)$, $B'(4, 5)$
    (c) Coordinates of endpoints are: $A''(-4, -1)$, $B''(-1, 0)$
    (d) Coordinates of endpoints are: $A'''(4, 1)$, $B'''(1, 0)$
    (e) $T_{0, -4}(x, y)$
13. **(b)** Coordinates of vertices are: $A'(1, 3)$, $B'(3, 4)$, $C'(1, 6)$
    (c) Coordinates of vertices are: $A''(-3, 1)$, $B''(-4, 3)$, $C''(-6, 1)$
    (d) Coordinates of vertices are: $A'''(3, -1)$, $B'''(4, -3)$, $C'''(6, -1)$

Section 4.6
1. **(a)** $(-7, 4)$ **(b)** $(2, 5)$ **(c)** $(-3, 0)$ **(d)** $(0, -4)$ **(e)** $(3, -2)$
2. **(a)** $(-2, 2)$ **(b)** $(0, 3)$ **(c)** $(-5, 3)$ **(d)** $(2, -3)$ **(e)** $(3, 6)$
3. **(a)** $(-1, 10)$ **(c)** $(2, 5)$ **(e)** $(4, -6)$
    (b) $(-3, 4)$ **(d)** $(0, -1)$ **(f)** $(-5, -2)$
4. $h = 1, k = 5$ **5.** (2) **6.** (2) **7.** (3)
8. **(a)** $\overline{AH}$ **(b)** $\overline{AH}$ **(c)** G
9. **(a)** $(3, 5)$ **(b)** $(1, -1)$ **(c)** $(5, 5)$
10. **(b)** Coordinates of vertices are: $A'(1, 6)$, $B'(5, 8)$, $C'(4, 5)$
    (c) Coordinates of vertices are: $A''(-6, -1)$, $B''(-8, -5)$, $C''(-5, -4)$
    (d) Coordinates of vertices are: $A'''(9, 0)$, $B'''(11, 4)$, $C'''(8, 3)$
11. **(b)** Coordinates of vertices are: $A'(0, 4)$, $B'(-1, 8)$, $C'(-4, 8)$
    (c) (1)
    (d) Coordinates of vertices are: $A''(4, -8)$, $B''(8, -7)$, $C''(8, -4)$
    (e) (4)

Section 4.7
1. Does not exist since the inverse relation is not a function.
2. $g^{-1} = \{5, -2), (-2, 5),(2, 1)\}$ **5.** $y = \dfrac{x}{3} - \dfrac{2}{3}$
3. $y = 3 - x$ **6.** $y = \sqrt{x^2 - 1}$
4. $y = \sqrt[3]{x - 1}$ **7.** $y = -\dfrac{3}{2}x + \dfrac{5}{2}$
8. Does not exist since the inverse relation, $y = \pm\sqrt{2x - 8}$, is not a function.
9. Show that $f(g(x)) = x$ and $g(f(x)) = x$.
11. Reflection in the line $y = x$.
13. Fails the horizontal-line test, so no inverse function exists.
14. Passes the horizontal-line test, so an inverse function exists.

Section 4.8
1. 9   **2.** 4   **3.** 14   **4.** 16
5. (b) $xy = 3$ **(c)** $xy = -12$
6. (b) $xy = 10$
7. Eight

Regents Tune-Up: Chapter 4

1. (4, 7)	**7.** C	**13.** (1)
2. 81	**8.** $\overline{AE}$	**14.** (3)
3. (2, 8)	**9.** C	**15.** (1)
4. 8	**10.** 70	**16.** (3)
5. –1	**11.** (0, –3)	**17.** (3)
6. (0, –4)	**12.** 13	

18. (a) $A'(0, 3), B'(6, -4)$
    (b) $A'(0, 3), B'(-6, -4)$
    (c) $A'(1, 0), B'(8, 6)$
    (d) $A'(-3, 0), B'(4, -6)$
    (e) $A'\left(\dfrac{3}{2}, 0\right), B'(-2, 3)$

19. (b) Coordinates of vertices are: $A'(4, -1), B'(7, 3), C'(1, 5)$
    (c) Reflection in the line $y = x$

CHAPTER 5

EXPONENTIAL AND LOGARITHMIC FUNCTIONS

5.1 GRAPHING $y = b^x$ AND ITS INVERSE

=== KEY IDEAS ===

An **exponential function** raises a positive constant other than 1 to a variable power. The function $y = 2^x$ is an exponential function. Its inverse is $x = 2^y$.

In order to solve for y in terms of x, the **logarithm function** was invented. If $x = 2^y$, then $y = \log_2 x$, which is read as "logarithm of x to base 2 is y." A logarithm is an *exponent* since, if $y = \log_b x$, then $x = b^y$.

Graphing $y = b^x$ ($b > 0$)

In Figure 5.1, the graphs of $y = 2^x$, $y = 10^x$, and $y = \left(\frac{1}{2}\right)^x$ are drawn on the same set of axes. The graph of $y = \left(\frac{1}{2}\right)^x$ is a reflection of the graph of $y = 2^x$ in the y-axis, since replacing x by $-x$ in $y = 2^x$ gives

$$y = 2^{-x} = (2^{-1})^x = \left(\frac{1}{2}\right)^x.$$

Thus, graphs of the forms

$$y = b^x \quad \text{and} \quad y = \left(\frac{1}{b}\right)^x$$

are reflections of each other in the y-axis. Furthermore, the graph of any exponential function $y = b^x$, where $b > 0$ and $b \neq 1$, has the x-axis as an asymptote, lies in Quadrants I

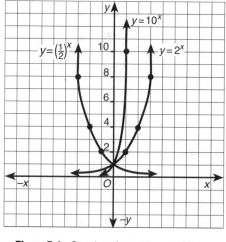

Figure 5.1 Graphs of $y = 2^x$, $y = 10^x$ and $y = \left(\frac{1}{2}\right)^x$

and II, and contains point $(0, 1)$. If $0 < b < 1$, the graph falls. If $b > 1$, the graph rises, with its steepness increasing as b gets larger. Notice, for example, that the graph of $y = 10^x$ in Figure 5.1 is steeper than the graph of $y = 2^x$.

The graph passes the vertical- and horizontal-line tests, so $y = b^x$ is a one-to-one function.

Graphing the Inverse of $y = b^x$ ($b > 0$ and $b \neq 1$)

The inverse of the exponential function $y = b^x$ is $x = b^y$, which may also be written as $y = \log_b x$, where "log" is an abbreviation for *logarithm*.

MATH FACTS

DEFINITION OF LOGARITHM

The **logarithm** of a positive number x to a given base b is the power to which b must be raised to obtain x, provided that $b > 0$ and $b \neq 1$. Thus

$$y = \log_b x \quad \text{means} \quad b^y = x.$$

The graphs of the exponential and logarithmic functions are shown in Figure 5.2. Since $y = \log_b x$ is the inverse of $y = b^x$, either graph is a reflection of the other in the line $y = x$. The key features of these graphs are compared in Table 5.1. Notice that the graph of every logarithmic function passes through $(1, 0)$, and the graph of every exponential function passes through $(0, 1)$.

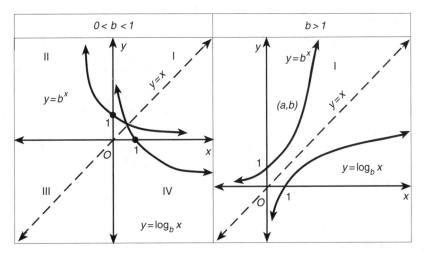

Figure 5.2 Graphs of $y = b^x$ and $y = \log_b x$

TABLE 5.1 COMPARISON OF THE FEATURES OF THE GRAPHS OF
$y = b^x$ **AND** $y = \log_b x$

Feature	Graph of $y = b^x$	Graph of $y = \log_b x$
Quadrants where located	I and II	I and IV
Domain	All real numbers	Positive real numbers
Range	Positive real numbers	All real numbers
x-intercept	None	(1, 0)
y-intercept	(0, 1)	None
Asymptote	x-axis	y-axis
$0 < b < 1$	Falls	Falls
$b > 1$	Rises	Rises

Example

1. On the same set of axes, sketch and label the graphs of:
(a) $y = 3^x$ and its inverse
(b) the reflection of $y = 3^x$ in the y-axis

Solution: **(a)** To sketch the graph of $y = 3x$, use the sample points shown in the table at the right. The graph is shown below.

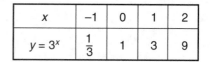

x	−1	0	1	2
$y = 3^x$	$\frac{1}{3}$	1	3	9

The inverse of $y = 3^x$ is the logarithmic function whose equation is $y = \log_3 x$. To obtain some sample points, interchange the x- and y-values in the given table. Notice in the accompanying figure that the graph of $y = \log_3 x$ is the reflection of the graph of $y = 3^x$ in the line $y = x$.

(b) In general, the graphs of $y = b^x$ and $y = \left(\frac{1}{b}\right)^x$ are reflections of each other in the y-axis. In this example, $b = 3$ so that $y = \left(\frac{1}{3}\right)^x$ is the reflection of $y = 3^x$ in the y-axis. Some sample points are given in the accompanying table. See the figure for part **(a)**.

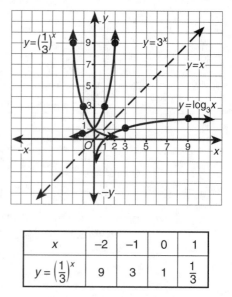

x	−2	−1	0	1
$y = \left(\frac{1}{3}\right)^x$	9	3	1	$\frac{1}{3}$

106

Equivalent Exponential and Logarithmic Forms

The fact that $y = \log_b x$ means $x = b^y$ makes it possible to write equivalent exponential and logarithmic equations, as illustrated in Table 5.2.

TABLE 5.2 EQUIVALENT EXPONENTIAL AND LOGARITHMIC EQUATIONS

Exponential Form	Logarithmic Form
$4^3 = 64$	$\log_4 64 = 3$
$10^2 = 100$	$\log_{10} 100 = 2$
$5^0 = 1$	$\log_5 1 = 0$
$3^{-2} = \dfrac{1}{9}$	$\log_3\left(\dfrac{1}{9}\right) = -2$

Examples

2. For each logarithmic equation, find x:
(a) $\log_2 x = 5$ **(b)** $\log_7 49 = x$ **(c)** $\log_x 8 = \dfrac{3}{4}$

Solutions: In each case change to exponential form.
(a) If $\log_2 x = 5$, then $2^5 = x$. Hence $x = \textbf{32}$.
(b) If $\log_7 49 = x$, then $7^x = 49$. Since $7^x = 49 = 7^2$, $x = \textbf{2}$.
(c) If $\log_x 8 = \frac{3}{4}$, then $x^{\frac{3}{4}} = 8$. Solve for x by raising both members of the equation to the $\frac{4}{3}$ power.

$$x = 8^{\frac{4}{3}} = \left(\sqrt[3]{8}\right)^4 = (2)^4 = \textbf{16}.$$

3. If $\log_x 5 = 3$, solve for x correct to the *nearest tenth*.

Solution: If $\log_x 5 = 3$, then $x^3 = 5$, so $x = 5^{\frac{1}{3}}$. Using your calculator, press

$$5 \boxed{x^y} \boxed{(} 1 \boxed{\div} 3 \boxed{)} \boxed{=} .$$

Thus, correct to the *nearest tenth*, x is **1.7**.

Common Logarithms

A logarithm whose base is 10 is called a **common logarithm**. A common logarithm is indicated when the base is omitted. Thus, $\log x$ is understood to mean $\log_{10} x$.

The common logarithm of any power of 10 can be determined by sight. Thus, $\log 10 = 1$ since $10^1 = 10$; $\log 100 = 2$ since $10^2 = 100$; $\log 1000 = 3$ since $10^3 = 1000$; and so forth.

Exercise Set 5.1

1–4. Change from exponential into logarithmic form.

1. $2^4 = 16$ **2.** $8^{\frac{2}{3}} = 4$ **3.** $6^0 = 1$ **4.** $5^{-2} = \dfrac{1}{25}$

5–13. Change from logarithmic into exponential form, and find x.

5. $\log_6 36 = x$ **8.** $\log_5 \dfrac{1}{5} = x$ **11.** $\log_x 16 = \dfrac{4}{3}$

6. $\log x = -1$ **9.** $\log_9 x = \dfrac{3}{2}$ **12.** $\log_x 16 = \dfrac{2}{3}$

7. $\log_x 0.01 = -2$ **10.** $\log_3(2 - x) = 2$ **13.** $\log_x 8 = -\dfrac{3}{2}$

14. If $f(x) = \log x$, find:
    (a) $f(10)$ **(b)** $f(100)$ **(c)** $f(1000)$ **(d)** $f(0.1)$

15. Find the value of k if the following points lie on the graph of $y = \log_{10} x$:
    (a) $(k, 0)$ **(b)** $(10, k)$ **(c)** $(k, -1)$

16. At which point do the graphs of $y = 5^x$ and $y = 5^{-x}$ intersect?

17. At which point do the graphs of $y = \log_4 x$ and $y = \log_{\frac{1}{4}} x$ intersect?

18. Sketch the graph of $y = 2^x$. On the same set of axes, sketch the reflection of $y = 2^x$ in:
    (a) the y-axis   **(b)** the line $y = x$

19. Solve for x correct to the *nearest tenth.*
    (a) $\log_x 17 = 5$ **(b)** $\log_x 7 = \dfrac{3}{4}$ **(c)** $\log_3 x = \dfrac{8}{5}$

20. Graph and label the function $y = 2^x$ for the domain $-2 \le x \le 3$. On the same set of axes, sketch:
    (a) the reflection in the x-axis of $y = 2^x$, and label the graph **(a)**
    (b) the reflection in the origin of $y = 2^x$, and label the graph **(b)**
    (c) the translation of $y = 2^x$ under $T_{(6,2)}$, and label the graph **(c)**

21. On the same set of axes, sketch the graphs of $y = \left(\dfrac{1}{2}\right)^x$ and $y = \log_{\frac{1}{2}} x$. Label each graph with its equation.

5.2   USING LOGARITHM LAWS

Since logarithms are exponents, the laws for finding the logarithms of a product, quotient, and power are consistent with the laws of exponents for these operations.

Power Property	Logarithm Property
$b^m \cdot b^n = b^{m+n}$	$\log_b(xy) = \log_b x + \log_b y$
$\dfrac{b^m}{b^n} = b^{m-n}$	$\log_b\left(\dfrac{x}{y}\right) = \log_b x - \log_b y$
$(b^m)^n = b^{mn}$	$\log_b(x^n) = n \log_b x$

Properties of Logarithms

Common logarithms of products, quotients, powers, and roots can be simplified by using one or more of the logarithm laws summarized in Table 5.3. These laws are also true when the base of the logarithm is a positive number other than 10.

TABLE 5.3 LOGARITHM LAWS

Logarithm Law	Explanation	General Rule
Product Law	The logarithm of a product equals the sum of the logarithms of its factors. For example: $\log 14 = \log(7 \times 2) = \textbf{log 7} + \textbf{log 2}$	$\log(xy) = \log x + \log y$
Quotient Law	The logarithm of a quotient equals the logarithm of the numerator minus the logarithm of the denominator. For example: $\log 3.5 = \log\left(\dfrac{7}{2}\right) = \textbf{log 7} - \textbf{log 2}.$	$\log\left(\dfrac{x}{v}\right) = \log x - \log y$
Power Law	The logarithm of a quantity raised to a power equals the exponent times the logarithm of that quantity. For example: $\log 49 = \log(7^2) = \textbf{2 log 7}.$	$\log(x^n) = n \cdot \log x$

Here are some additional examples:

- $\log 100x = \log 100 + \log x = 2 + \log x$

- $\log\left(\dfrac{x}{10}\right) = \log x - \log 10 = \log x - 1$

- $\log\left(\dfrac{p}{p^2}\right) = \log p - \log(q^2) = \log p - 2\log q$

- $\log\left(\dfrac{\sqrt{x}}{y}\right) = \log\left(\dfrac{x^{\frac{1}{2}}}{y}\right) = \dfrac{1}{2}\log x - \log y$

Rewriting Logarithmic Expressions

To rewrite a logarithmic expression such as $\frac{1}{2}(\log r - 3\log s)$ in terms of the logarithm of a single term or an algebraic expression, undo one or more logarithm laws, beginning with the expression inside the parentheses:

- Undo the Power Law inside the parentheses:

$$\frac{1}{2}(\log r - 3\log s) = \frac{1}{2}(\log r - \log s^3)$$

- Undo the Quotient Law inside the parentheses:

$$= \frac{1}{2}\log\left(\frac{r}{s^3}\right)$$

- Undo the Power Law:

$$= \log\left(\frac{r}{s^3}\right)^{\frac{1}{2}}$$

$$= \log\sqrt{\frac{r}{s^3}}$$

Examples

1. If $\log x = n$, express $\log\left(\dfrac{x^3}{100}\right)$ in terms of n.

Solution:

Use Quotient Law: $\qquad \log\left(\dfrac{x^3}{100}\right) = \log x^3 - \log 100$

Use Power Law: $\qquad\qquad\qquad = 3\log x - \log 100$
Simplify common log: $\qquad\qquad = 3\log x - 2$
Replace $\log x$ by n: $\qquad\qquad = 3n - 2$

2. If $\log 2 = x$ and $\log 3 = y$, express each of the following in terms of x and y:

(a) $\log \sqrt{6}$ **(b)** $\log 24$ **(c)** $\log 1.5$

Solution: **(a)** $\qquad\qquad\qquad\qquad \log \sqrt{6} = \log 6^{\frac{1}{2}}$

Use Power Law: $\qquad\qquad\qquad\qquad = \dfrac{1}{2} \log 6$

Factor 6: $\qquad\qquad\qquad\qquad\quad = \dfrac{1}{2} \log(2.3)$

Use Product Law: $\qquad\qquad\qquad\quad = \dfrac{1}{2}(\log 2 + \log 3)$

Let $\log 2 = x$ and $\log 3 = y$: $\qquad = \dfrac{1}{2}(x + y)$

(b) Since 24 can be factored as $8 \cdot 3$ or $2^3 \cdot 3$,

$$\log 24 = \log(2^3 \cdot 3)$$

Use Product Law: $\qquad = \log 2^3 + \log 3$

Use Power Law: $\qquad = 3 \log 2 + \log 3$

$$= 3x + y$$

(c) $\log 1.5 = \log\left(\dfrac{3}{2}\right)$

$$= \log 3 - \log 2$$
$$= y - x$$

3. If $\log A = 0.4262$, find $\log \sqrt{A}$.

Solution: Use the Power Law:

$$\log \sqrt{A} = \log A^{\frac{1}{2}} = \dfrac{1}{2} \log A = \dfrac{1}{2}(0.4262) = \mathbf{0.2131}$$

4. If $x = \dfrac{a\sqrt{b}}{c}$, express $\log x$ in terms of $\log a$, $\log b$, and $\log c$.

Solution: $\qquad\qquad \log x = \log\left(\dfrac{a\sqrt{b}}{c}\right)$

Use Quotient Law: $\quad = \log\left(a\sqrt{b}\right) \qquad - \log c$

Use Product Law: $\quad = \log a + \log \sqrt{b} - \log c$

Use Power Law: $\qquad = \log a + \log b^{\frac{1}{2}} - \log c$

$$= \mathbf{\log a + \dfrac{1}{2} \log b - \log}$$

111

5. If $\log N = 2 \log x + \log y$, express N in terms of x and y.

Solution: Write the right side of the equation as the logarithm of a single expression:

$$\begin{aligned} \log N &= 2 \log x + \log y \\ &= \log x^2 + \log y \\ &= \log x^2 y \end{aligned}$$

If the *logarithms* of two expressions are equal, then the two expressions are equal, provided that the bases are the same. Thus

$$N = x^2 y.$$

Solving Logarithmic Equations

An equation in which each term is a logarithm is solved by putting the equation in the form

$$\log_b N = \log_b M.$$

If $\log_b N = \log_b M$, then $N = M$ since the bases are the same.

Example

6. Solve for x: $\log x - \dfrac{1}{3} \log 8 = \log 7$.

Solution:

Use Power Law:	$\log x - \log 8^{\frac{1}{3}} = \log 7$
Simplify:	$\log x - \log 2 = \log 7$
Use Quotient Law:	$\log \dfrac{x}{2} = \log 7$
Equate numbers since their logs are equal:	$\dfrac{x}{2} = 7$

The check is left for you.

Some logarithmic equations are solved by forming and then solving the equivalent exponential equation.

Example

7. Find x if $\log_4 (x - 3) + \log_4 (x + 3) = 2$.

Solution:

Use Product Law:	$\log_4 [(x - 3)(x + 3)] = 2$
Put into exponential form:	$(x - 3)(x + 3) = 4^2$
Multiply:	$x^2 - 9 = 16$

Take square roots:
$$x^2 = 25$$
$$x = 5 \text{ or } x = -5$$

Checking the root $x = -5$ in the original equation leads to the sum $\log_4(-8)$ + $\log_4(-2)$ on the left side of the equation. Since logarithms of negative numbers are undefined, reject $x = -5$. The solution is **$x = 5$**. The check for this root is left for you.

Exercise Set 5.2

1–4. If log x = n, *express in terms of* n:

1. $\log \dfrac{x^2}{10}$ **2.** $\log \dfrac{100}{\sqrt{x}}$ **3.** $\log x\sqrt{x}$ **4.** $\log \sqrt{1000x}$

5. The expression $\log \dfrac{\sqrt{x^2 y^3}}{z}$ is equivalent to

(1) $\dfrac{1}{2}(2 \log x + 3 \log y - \log z)$

(2) $\dfrac{1}{2}(2 \log x + 3 \log y) - \log z$

(3) $2 \log x + 3 \log y - \log z$

(4) $\dfrac{x^2 y^3}{z}$

6. If $\log 3 = x$ and $\log 5 = y$, express in terms of x and y:
(a) $\log 15$ (b) $\log 45$ (c) $\log 75$ (d) $\log 0.6$ (e) $\log \dfrac{5}{9}$

7. If $\log 2 = 0.3010$ and $\log 7 = 0.8451$, find:
(a) $\log 49$ (b) $\log 3.5$ (c) $\log 70$ (d) $\log 28$ (e) $\log \sqrt{2}$

8–13. Express log x *in terms of log* m, *log* n, *and log* p.

8. $x = \sqrt{mnp}$ **10.** $x = \dfrac{p}{\sqrt{mn}}$ **12.** $x = \dfrac{n\sqrt{m}}{p}$

9. $x = \dfrac{mp}{n^2}$ **11.** $x = \dfrac{np}{\sqrt{m}}$ **13.** $x = \dfrac{(mp)^2}{n}$

14. If $V = \pi r^2 h$, write $\log V$ in terms of $\log \pi$, $\log r$, and $\log h$.

15–20. Write each logarithmic expression as the logarithm of a single expression.

15. $2 \log r - \dfrac{1}{3} \log s$

16. $\dfrac{1}{2} \log r + 3 \log s$

17. $\log r + 2 \log s - \log t$

18. $\dfrac{1}{3}(\log r - 2 \log t) + \log s$

19. $3\left(\log s + \dfrac{1}{2} \log t\right) - \log r$

20. $2 \log t - \left(\log s + \dfrac{1}{3} \log r\right)$

21 and 22. Express N in terms of x and y.

21. $\log N = 2 \log x - 3 \log y$

22. $\log N = x \log y + 2 \log y$

23–30. Solve for x:

23. $2 \log_5 x = \log_5 16$

24. $\log_x 4 + \log_x 16 = 2$

25. $\log_3(7x + 4) - \log_3 2 = 2 \log_3 x$

26. $\log_x 2 + \log_x 4 = \dfrac{3}{2}$

27. $\log x + \dfrac{1}{2} \log 25 = 2 \log 10$

28. $\log_3(x - 4) + \log_3(x + 4) = 2$

29. $\log_6(x + 5) + \log_6 x = 2$

30. $\log_9 x + \log_9 (x - 6) = \dfrac{3}{2}$

5.3 COMPUTING WITH LOGARITHMS

KEY IDEAS

Before computers and calculators became widely available, common logarithm values were obtained from tables and then used to perform complicated arithmetic calculations applying the properties of logarithms. Instead of working with tables, a scientific calculator can be used to find a common logarithm of any positive number.

Finding Common Logarithms Using a Calculator

To find the value of log x, enter x and then press the $\boxed{\text{LOG}}$ key. Some calculators require you to press the $\boxed{\text{LOG}}$ key before entering the number. To evaluate log 100, key in

$$\boxed{\text{LOG}} \ 100 \ \boxed{=} \quad \text{or} \quad 100 \ \boxed{\text{LOG}}.$$

Usually, the logarithm of a number that is not a power of 10 must be rounded off to four decimal places. Use your calculator to verify that log 24, correct to four decimal places, is 1.3802. In other words, log 24 ≈ 1.3802, where the symbol ≈ is read as "is approximately equal to."

Logarithms of Numbers Between 0 and 1

The graph of $y = \log_b x$ when $b > 1$ in Figure 5.2 on page 105 illustrates that, for all x-values between 0 and 1, the curve falls below the x-axis, so the corresponding y-values are negative. This means that the common logarithm of a number between 0 and 1 will always be a negative number. Use your calculator to verify that log 0.01 = –2, meaning that point (0.01, –2) is on the graph of $y = \log x$. As another example, use your calculator to verify that log 0.0326 ≈ –1.4868.

Finding Antilogarithms of Numbers

If you know the logarithm of a number N (the exponent of 10 that produces N), you can use your scientific calculator to find N by finding the **antilogarithm** of the exponent. Thus, if log $N = 1.7657$, then

$$N = 10^{1.7657} = \text{antilogarithm of } 1.7657.$$

Since finding logarithms and finding antilogarithms are inverse operations, use the key on your scientific calculator labeled 10^x, the inverse of log x, to find an antilogarithm. For example, to find the antilogarithm of 1.7657 correct to the *nearest hundredth*:

- Locate 10^x on your calculator. Since 10^x is usually printed directly *above* the ⌊LOG⌋ key, you will probably have to press the ⌊2nd⌋ function key before pressing the key that accesses the 10^x function.
- Use 1.7657 for the value of x and one of the following key sequences:

  ⌊2nd⌋ 10^x 1.7657 ⌊=⌋   or   1.7657 ⌊2nd⌋ 10^x.

- Round off the displayed value to the required level of accuracy. Thus, the antilogarithm of 1.7657, correct to the *nearest hundredth*, is 58.3.

MATH FACTS

FINDING ANTILOGARITHMS

If log $N = k$, then N = antilogarithm of k. To find N when x is given, key in

⌊2nd⌋ ⌊10^x⌋ k ⌊=⌋   or k ⌊2nd⌋ ⌊10^x⌋.

Examples

1. If log $N = -1.1302$, find N correct to the *nearest one-thousandth*.

Solution: Since the logarithm value (exponent of 10) is a negative number, enter it *without* its sign and then press the change-of-sign key, $\boxed{+/-}$, as in

$$\boxed{\text{2nd}} \ \boxed{10^x} \ 1.1302 \ \boxed{+/-} \ \boxed{=} \quad \text{or} \quad 1.1302 \ \boxed{+/-} \ \boxed{\text{2nd}} \ \boxed{10^x}.$$

Since $N \approx 0.074096893$, N correct to the *nearest one-thousandth* is **0.074**.

2. If $\log\left(\dfrac{x}{4}\right) = 1.5$, find x correct to the *nearest tenth*.

Solution: If $\log\left(\dfrac{x}{4}\right) = 1.5$, then $\dfrac{x}{4} = $ antilogarithm of 1.5, so

$$x = 4 \times \text{antilogarithm of } 1.5.$$

Using your scientific calculator, key in

$$4 \ \boxed{\times} \ \boxed{\text{2nd}} \ \boxed{10^x} \ 1.5 \ \boxed{=} \quad \text{or} \quad 4 \ \boxed{\times} \ 1.5 \ \boxed{\text{2nd}} \ \boxed{10^x} \ \boxed{=}.$$

Since $x \approx 126.4911$, the value of x correct to the *nearest tenth* is **126.5**.

Exercise Set 5.3

1. Given log $2.45 = 0.3892$, find:
   (a) log 245 **(b)** log 0.0245 **(c)** $\log(2.45)^3$ **(d)** $\log \sqrt[3]{0.245}$

2–4. Estimate x *to the* nearest hundredth.

2. log $x = 0.8131$ **3.** log $x = 0.6100$ **4.** log $x = 0.7040$

5. Find the integer whose logarithm is between 1.5611 and 1.5466.

6. Find the number correct to the *nearest hundredth* whose logarithm is closest to:
   (a) -0.8041 **(b)** -1.4949

7. If log $100x = 1.3488$, find x correct to the *nearest hundredth*.

8. If $(\log x)^2 = 3.7812$, find x correct to the *nearest hundredth*.

9. If $\log \sqrt{x} = -0.6419$, find x correct to the *nearest hundreth*.

10. If $\log(x - 3) = 1.0317$, find x correct to the *nearest tenth*.

11. If log $N = \dfrac{\log 13.4}{\log 100}$, find N correct to the *nearest tenth*.

12. If $\log\left(\dfrac{1}{N}\right) = \dfrac{1}{\log 56.2}$, find N correct to the *nearest hundredth*.

5.4 SOLVING EXPONENTIAL EQUATIONS

Equations such as

$$5^{x+1} = 25^x, \quad 3^{x-4} = 9, \quad \text{and} \quad 2^{3x} = 6$$

are called *exponential* equations. An **exponential equation** is an equation in which the variable is contained in one or more exponents. Some exponential equations can be solved by writing each side of the equation as a power of the same base and then equating the exponents. For example, if $3^{x-4} = 9$, then $3^{x-4} = 3^2$, implying that the exponents $x - 4$ and 2 are equal so that $x - 4 = 2$ or $x = 6$.

When both sides of an exponential equation cannot be expressed as a power of the same base, as in $2^{3x} = 6$, logarithms are needed.

Exponential Equations Having a Common Base

If $b^m = b^n$, then $m = n$ (provided that $b \neq 1$). This principle is used to solve exponential equations in which both sides can be expressed as powers of the same base.

Example

1. Solve for x:
(a) $9^{x+1} = 27x$ (b) $64^{1-x} = \dfrac{1}{16}$

Solutions:

(a)

	$9^{x+1} = 27^x$
Write each side as a power of 3:	$(3^2)^{x+1} = (3^3)^x$
Use Power Law of Exponents:	$3^{2(x+1)} = 3^{3x}$
Equate exponents:	$2(x + 1) = 3x$
Solve for x:	$2x + 2 = 3x$
	$x = 2$

(b)

	$64^{1-x} = \dfrac{1}{16}$
Write each side as a power of 4:	$(4^3)^{1-x} = 4^{-2}$
Use Power Law of Exponents:	$4^{3(1-x)} = 4^{-2}$
Equate exponents:	$3(1 - x) = -2$
Solve for x:	$3 - 3x = -2$
	$-3x = -5$
	$x = \dfrac{5}{3}$

Exponential Equations Lacking a Common Base

To solve an exponential equation for which it is not possible to express both sides as rational powers of the same base, take the logarithm of each side of the equation. In general, if $a^x = b$, then

$$x \log a = \log b \quad \text{so that } x = \frac{\log b}{\log a}.$$

Examples

2. Solve the equation $2^{3x} = 6$ for x to the *nearest hundredth*.

Solution: Solve by using logarithms.

Take log of each side:	$\log (2^{3x}) = \log 6$
Use Power Law of Logs:	$3x \log 2 = \log 6$
Solve for x:	$x = \frac{1}{3}\left(\frac{\log 6}{\log 2}\right)$
Evaluate common logs:	$= \frac{1}{3}\left(\frac{0.7782}{0.3010}\right)$
Divide:	$= \frac{0.7782}{0.9030} = 0.8618$
Round to the nearest hundredth.	$= \mathbf{0.86}$

3. Solve the equation $\log_3 21 = x$ to the *nearest tenth*.

Solution: If $\log_3 21 = x$, then $3^x = 21$.

Take log of each side:	$x \log 3 = \log 21$
Solve for x:	$x = \frac{\log 21}{\log 3}$
Evaluate common logs:	$= \frac{1.3222}{0.4771}$
Divide:	$= 2.77$
Round to the nearest tenth:	$= \mathbf{2.8}$

Exercise Set 5.4

1–9. Solve each equation for x.

1. $2^{x-3} = 64$

2. $4^{2x+1} = 8^x$

3. $6^{1-4x} = \frac{1}{6}$

4. $27^x = 3^{2x-1}$

5. $64 = 16^{7x-2}$

6. $2^{1-3x} - \frac{1}{4} = 0$

7. $9^{2x} = \frac{1}{81}$

8. $\left(\frac{1}{8}\right)^{2x+1} = 4$

9. $7^{x^2-2x} = 1$

10–15. Using logarithms, solve each equation for x to the nearest tenth.

10. $4^x = 12$ **12.** $\left(\dfrac{1}{3}\right)^{-x} = 7$ **14.** $\log_6 10 = x$

11. $3^{2x} = 75$ **13.** $27^x = 500$ **15.** $3\log_3 2 = x$

16. The depreciation (decline in cash value) on a car can be determined by the formula $V = C(1 - r)^t$, where V = value of the car after t years, C = original cost, and r = rate of depreciation. A car's cost, when new, is \$20,000. If the rate of depreciation is 30% and the current value of the car is \$4000, how old, to the *nearest tenth of a year*, is the car?

17. The amount, A, in milligrams, of a 10-milligram dose of a drug remaining in the body after t hours is given by the formula $A = 10(0.8)^t$. Find, to the *nearest tenth of an hour*, how much time has passed when half of the drug dose is left in the body.

REGENTS TUNE-UP: CHAPTER 5

Each of the questions in this section has appeared on a previous Course III Regents Examination. Here is an opportunity for you to review Chapter 5 and, at the same time, prepare for the Course III Regents Examination.

---

1. The graph of the function $f(x) = 3^x$ lies in which quadrant(s)?

2. Find the value of x: $\log_3(x - 2) = 2$.

3. If $\log_x\left(\dfrac{1}{4}\right) = -1$, find x.

4. Using logarithms, solve the equation $(1.95)^x = 54$ for x to the *nearest integer*.

5. If $\log_3 5 = x$, find x to the *nearest tenth*.

6. If $x = \dfrac{\sqrt{r}}{s}$, which expression is equivalent to $\log x$?

   (1) $2 \log r$ (3) $\dfrac{1}{2} \log r - \log s$

   (2) $2 \log r - \log s$ (4) $\dfrac{\log r - \log s}{2}$

7. If $\log 3 = a$ and $\log 5 = b$, then $\log 45$ is equal to:
   (1) $a^2 + b$ (2) $2a + b$ (3) $2ab$ (4) $a^2 b$

8. Solve for x: $4^{3x} = 8^{x+1}$.

9. The expression $3 \log x - \dfrac{1}{2} \log y$ is equal to:

(1) $\log\left(\dfrac{x^3}{y^2}\right)$ (2) $\log\left(\dfrac{x^3}{\sqrt{y}}\right)$ (3) $\log\sqrt{\dfrac{3x}{y}}$ (4) $\dfrac{\log 3x}{\frac{1}{2}\log y}$

10. If $\log_9 x = \dfrac{3}{2}$, what is the value of x?

(1) $\dfrac{3}{2}$ (2) 8   (3) $\dfrac{27}{2}$ (4) 27

11. (a) Solve for x, and express the roots in terms of i: $2x^2 = 6x - 5$.
  (b) Using logarithms, find x to the *nearest tenth*: $3^{2x} = 100$.

12. Given: $f(x) = 2^x$.
  (a) On graph paper, sketch the graph of $f(x)$ in the interval $-1 \le x \le 3$. Label the graph **(a)**.
  (b) On the same set of axes, sketch the image of the graph drawn in part **(a)** after $(T_{2,-1} \circ r_{x\text{-axis}})$. Label the graph **(b)**.
  (c) On the same set of axes, sketch the image of the graph drawn in part **(b)** after D_{-1}. Label the graph **(c)**.

13. (a) Given: $\log_b 2 = 0.6931$ and $\log_b 3 = 1.0986$.
     Find: $\log_b \sqrt{12}$.
  (b) Solve for x: $\log_8 (x - 6) + \log_8 (x + 6) = 2$.

ANSWERS TO SELECTED EXERCISES: CHAPTER 5

Section 5.1
1. $\log_2 16 = 4$
3. $\log_8 4 = 0$
5. 2
7. 10
9. 27
11. 8
13. $\dfrac{1}{4}$
15. (a) 1
   (b) 1
   (c) 0.1
17. $(1, 0)$
19. (a) 1.8
   (b) 13.4
   (c) 5.8

21.

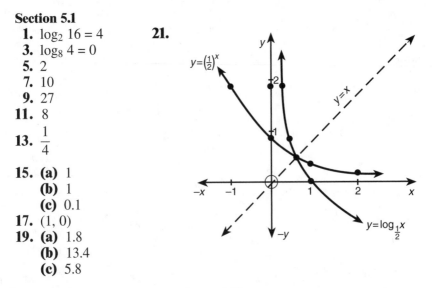

Section 5.2

1. $2n-1$

3. $\dfrac{3}{2}n$

5. (2)

6. (a) $x + y$
   (b) $2x + y$
   (c) $x + 2y$
   (d) $x - y$
   (e) $y - 2x$

7. (a) 1.6902
   (b) 0.4225
   (c) 1.8451
   (d) 1.4471
   (e) 0.1505

8. $\dfrac{1}{2}(\log m + \log n + \log p)$

9. $\log m + \log p - 2\log n$

11. $\log n + \log p - \dfrac{1}{2}\log m$

13. $2(\log m + \log p) - \log n$

15. $\dfrac{r^2}{\sqrt[3]{s}}$

16. $\dfrac{\sqrt{r}}{s^3}$

17. $\dfrac{rs^2}{t}$

18. $\sqrt[3]{\dfrac{r}{t^2}}\,s$

19. $\dfrac{\left(s\sqrt{t}\right)^3}{r}$

20. $\dfrac{t^2}{s\sqrt[3]{r}}$

21. $\dfrac{x^2}{y^3}$

22. y^{x+2}

23. 4

24. 8

25. 4

26. 4

27. 20

28. 5

29. 4

30. 9

Section 5.3

1. (a) 2.3892
   (b) −1.6108
   (c) 1.1676
   (d) 0.1297

2. 6.50
3. 4.07
4. 5.06
5. 36

6. (a) 0.16
   (b) 0.03
7. 0.22
8. 88.00 or 0.01

9. 0.05
10. 13.8
11. 3.7
12. 0.27

Section 5.4

1. 9

2. −2

3. $\dfrac{1}{2}$

4. −1

5. $\dfrac{1}{2}$

6. 1

7. −1

8. $-\dfrac{5}{6}$

9. 0 or 2

10. 1.8

11. 2.0

12. 1.8

13. 1.9

14. 1.3

15. 1.9

16. 4.5

17. 3.1

Regents Tune-Up: Chapter 5

1. Quadrants I and II

2. 11

3. 4

4. 6

5. 1.5

6. (3)

7. (2)

8. 1

9. (2)

10. (4)

11. (a) $\dfrac{3 \pm i}{2}$

(b) 2.1

13. (a) 1.2424

(b) 10

12.

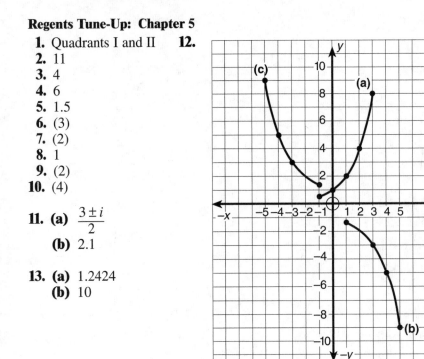

CHAPTER 6

CIRCLES

6.1 BEGINNING DEFINITIONS AND CIRCLE RELATIONSHIPS

==================== KEY IDEAS ====================

An important branch of geometry is concerned with measuring angles and special segments that intersect a circle. Before this subject can be discussed, however, some terms and relationships must be introduced. To simplify the discussion, the terms *equal* and *congruent* will be used interchangeably.

Special Segments and Lines of a Circle

A **circle** (Figure 6.1) is the set of all points that are the same distance from a fixed point in the plane, called the **center**. You should also know these facts:

- A **radius** of a circle is a segment whose endpoints are the center and a point on the circle. The plural of *radius* is *radii*. Radii of the same circle are congruent. Circles that have equal radii are congruent.

- A **chord** is a line segment whose endpoints are on the circle.

- A **diameter** is a chord that passes through the center of a circle. The length of a diameter is two times the length of a radius of the circle.

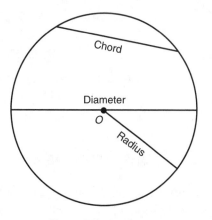

Figure 6.1 Circle *O*

- A **secant** is a line that intersects a circle in two different points. Every secant contains a chord of the circle. In Figure 6.2, $\overleftrightarrow{SEC}$ is a secant and $\overline{CES}$ is a *secant segment*. Since $\overline{CE}$ is the part of secant segment $\overline{CES}$ that lies outside the circle, it is called an *external secant segment*.

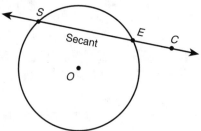

Figure 6.2 Secant and Secant Segment

- A **tangent** is a line that intersects a circle in exactly one point. The point of contact is called the *point of tangency*. In Figure 6.3, $\overleftrightarrow{TA}$ is a tangent and $\overline{TA}$ is a *tangent segment* since one of its endpoints, point A, is the point of tangency of $\overleftrightarrow{TA}$. Notice that radius $\overline{OA}$ is shown to be perpendicular to tangent $\overleftrightarrow{TA}$. This relationship between radii and tangents is always true.

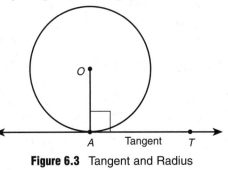

Figure 6.3 Tangent and Radius

MATH FACTS

RADIUS ⊥ TANGENT

A radius drawn to the point of tangency is perpendicular to the tangent at this point.

Congruent Triangles and Circles

Congruent triangles can be used to help establish important circle relationships. For example, in Figure 6.4, radii $\overline{OA}$ and $\overline{OB}$ are drawn to tangency points A and B. Right triangles OAP and OBP are congruent by the hypotenuse-leg theorem since $\overline{OP} \cong \overline{OP}$ (hypotenuse) and $\overline{OA} \cong \overline{OB}$ (leg).

Since corresponding parts of congruent triangles are equal, $PA = PB$. This proves that tangent segments drawn to a circle from the same point have the same length.

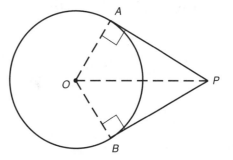

Figure 6.4 Equal Tangent Segments

===| **MATH FACTS** |===

EQUAL TANGENT SEGMENTS

If two tangent segments are drawn to a circle from the same point, they have the same length.

Example

1. In the accompanying diagram, $\overleftrightarrow{PAB}$ is tangent to circle R at A and tangent to circle S at B. $\overleftrightarrow{PCD}$ is tangent to circle R at C and tangent to circle S at D. Chord $\overline{BD}$ is drawn.
(a) If $m\angle PBD = 70$, find $m\angle P$.   **(b)** If $PA = 4$ and $PCD = 7$, find AB.

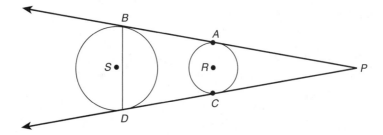

Solutions: Line segments PAB and PCD are drawn from the same point, P, and both are tangent to circle S, so they have the same length.
(a) Triangle BPD is isosceles with $PAB = PCD$. Since the measures of the base angles of an isosceles triangle are equal, $m\angle PDB = 70$. Hence

$$70 + 70 + m\angle P = 180$$
$$m\angle P = \mathbf{40}$$

(b) If $PCD = 7$, then $PAB = 7$. Since $AB + PA = PAB$ and $PA = 4$,

$$AB + 4 = 7$$
$$AB = \mathbf{3}$$

125

Polygons and Circles

If a polygon is **inscribed** in a circle, as in Figure 6.5, then each vertex of the polygon is a point on the circle. The circle is said to be *circumscribed* about the polygon.

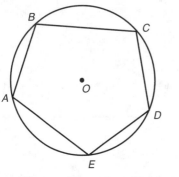

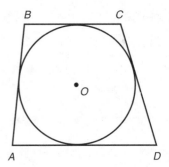

Figure 6.5 Polygon Inscribed in Circle

Figure 6.6 Polygon Circumscribed about Circle

If a polygon is **circumscribed** about a circle, as in Figure 6.6, then each of its sides is tangent to the circle. The circle is said to be *inscribed* in the polygon.

Arcs of a Circle

There are 360° in one complete revolution. Thus, the degree measure of a circle is 360. An **arc** of a circle is a curved portion of the circle whose degree measure is between 0 and 360.

- A **semicircle** is an arc whose endpoints are the endpoints of a diameter of the circle. The degree measure of a semicircle is 180.
- A **minor arc** is an arc whose degree measure is between 0 and 180. Minor arcs are named by their endpoints, as $\overarc{AB}$ in Figure 6.7.
- A **major arc** is an arc whose degree measure is between 180 and 360. Since, as in Figure 6.7, major arc AB and minor arc AB have the same endpoints, a major arc is named by using a third point on the circle, as $\overarc{APB}$. If $m\overarc{AB} = 70$, then

$$m\overarc{APB} = 360 - 70 = 290.$$

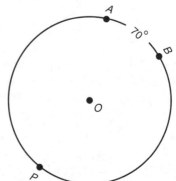

Figure 6.7 Major and Minor Arcs

In general, the degree measure of a major arc is 360 minus the measure of the minor arc that has the same endpoints.

- **Congruent arcs** are arcs in the same circle, or in congruent circles, that have the same degree measure.

Central Angles

Drawing two radii of a circle forms a *central* angle whose sides intercept a minor arc on the circle.

- A **central angle** is an angle whose vertex is at the center of a circle. In Figure 6.8, $\angle AOB$ is a central angle.
- The **degree measure of a minor arc** is the degree measure of the central angle that intercepts it. In Figure 6.8, $m\widehat{AB} = m\angle AOB$. If two central angles of a circle have the same measure, then their intercepted arcs will also have the same measure and will, therefore, be equal. Hence, if $m\angle AOB = m\angle COD$, then $m\widehat{AB} = m\widehat{CD}$.

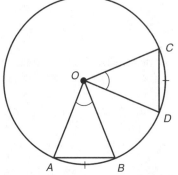

Figure 6.8 Measure of Minor Arc

Congruent Arcs and Chords

In Figure 6.9, if chord $\overline{AB}$ = chord $\overline{CD}$, then $m\widehat{AB} = m\widehat{CD}$. Conversely, if arcs AB and CD are equal, then chord $\overline{AB}$ = chord $\overline{CD}$. If $\overline{AB}$ is parallel to $\overline{CD}$, then $m\widehat{AB} = m\widehat{BD}$.

- Equal chords have equal arcs.
- Equal arcs have equal chords.
- Arcs between parallel chords are equal.

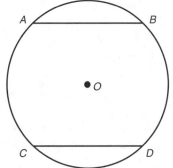

Figure 6.9 Congruent Arcs and Circles

Examples

2. In the accompanying diagram, $\overline{AB}$ is parallel to $\overline{CD}$. If $m\widehat{AB} = 110$ and $m\widehat{CD} = 90$, what is the measure of central angle *BOD*?

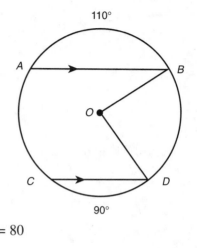

Solution: Since parallel chords intercept equal arcs, let $x = m\widehat{BD} = m\widehat{AC}$. The sum of the measures of the arcs that comprise a circle is 360. Thus

$$m\widehat{AB} + m\widehat{BD} + m\widehat{CD} + m\widehat{AC} = 360$$
$$110 + x + 90 + x = 360$$
$$2x + 200 = 360$$
$$x = \frac{160}{2} = 80$$

Hence, $m\widehat{BD} = 80$. Since a central angle and its intercepted arc have the same measure, $m\angle BOD = \mathbf{80}$.

3. In the accompanying diagram, regular hexagon *ABCDEF* is inscribed in circle *O*, and line *p* is a line of symmetry. What is the image of $r_p \circ R_{-240°}(F)$?

Solution: The given notation $r_p \circ R_{-240°}(F)$ means that point *F* must be rotated 240° in the *clockwise* direction, followed by a reflection of the image point in line *p*. In a *regular* hexagon, each of the six sides has the same length and, therefore, circle *O* is divided into six equal arcs.

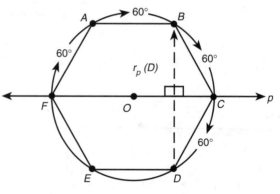

Since a circle measures 360°, the measure of each of the six arcs is $\frac{360°}{6}$, or 60°. Thus "moving" from one vertex to an adjacent vertex corresponds to a rotation of 60° about center *O*. Hence

$$r_p \circ R_{-240°}(F) = F \xrightarrow{R_{-240}} D \xrightarrow{r_p} \mathbf{B}.$$

4. In Example 3, does *ABCDEF* have *rotational* symmetry?

Solution: A figure has rotational symmetry if it is its own image under some rotation about its center. If *ABCDEF* is rotated 60° in either direction, as in the accompanying diagram, each vertex of the image will coincide with a vertex of the original figure. Hence a regular hexagon has **60° rotational symmetry**.

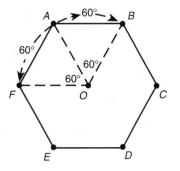

Diameters and Chords

In Figure 6.10, diameter $\overline{AOB}$ is drawn perpendicular to chord $\overline{CD}$ at *M*. As a result, chord segments $\overline{CM}$ and $\overline{DM}$ are congruent, as are the corresponding pairs of arcs on either side of the diameter. This can be proved by showing that right triangles *COM* and *DOM* are congruent.

Since $\overline{OC} \cong \overline{OD}$ (hyp) and $\overline{OM} \cong$ $\overline{OM}$ (leg), right triangles *COM* and *DOM* are congruent by the hyp-leg theorem. Thus

$$\overline{CM} \cong \overline{DM} \quad \text{and} \quad \angle 1 \cong \angle 2$$

since corresponding parts of congruent triangles are congruent (CPCTC). Since congruent central angles intercept congruent arcs, $\overset{\frown}{CB} \cong \overset{\frown}{DB}$. Subtracting these arcs from semicircles $\overset{\frown}{ACB}$ and $\overset{\frown}{ADB}$, respectively, gives $\overset{\frown}{AC} \cong \overset{\frown}{AD}$.

Referring to Figure 6.10, if it is known that $\overline{OB}$ bisects $\overline{CD}$, it may be concluded that $\overline{OB}$ is perpendicular to $\overline{CD}$.

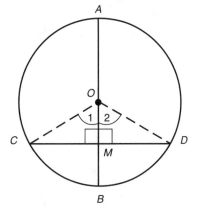

Figure 6.10 Diameters and Chords

DIAMETER ⊥ CHORD

If a diameter (or a line passing through the center of a circle) is perpendicular to a chord, then it bisects the chord and its arcs. Conversely, if a line passing through the center of a circle bisects a chord that is not a diameter, then it is perpendicular to the chord.

Equidistant Chords

In Figure 6.11, the lengths of perpendicular segments $\overline{OP}$ and $\overline{OQ}$ represent the distances between the center O and line segments $\overline{AB}$ and $\overline{CD}$, respectively. If $AB = CD$, then $OP = OQ$. Conversely, if $OP = OQ$, then $AB = CD$.

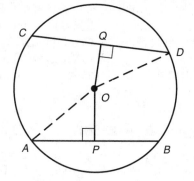

Figure 6.11 Equidistant Chords

EQUIDISTANT CHORDS

In a circle:

- Equal chords are the same distance from the center.
- Chords that are the same distance from the center are equal.

Example

5. A chord is 3 inches from the center of a circle whose radius is 5 inches. What is the length of the chord?

Solution: In the accompanying diagram, radius $\overline{OA}$ is 5 and *OE*, the distance of chord $\overline{AB}$ from center *O*, is 3. Thus, $\triangle OEA$ is a 3-4-5 right triangle, with *AE* = 4. Since $\overline{OE}$ is perpendicular to chord $\overline{AB}$ and passes through the center *O*, it lies on a diameter, so that it bisects the chord.

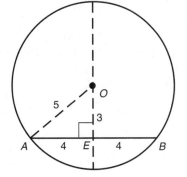

Hence *AE* = *BE* = 4.

The length of the chord is, therefore, 4 + 4 or **8** inches.

SUMMARY

1. In the same circle, two **arcs** are equal if any one of the following statements is true:

- The central angles that intercept the arcs are equal.
- The chords that the arcs determine are equal.
- The arcs are between parallel chords.
- The arcs are formed by a chord and a diameter drawn perpendicular to the chord.
- The arcs are semicircles.

2. In the same circle, two **chords** are equal if either of the following statements is true:

- The arcs that the chords intercept are equal.
- The chords are the same distance from the center.

Exercise Set 6.1

1–4. In the accompanying diagram, $\overline{AB}$ is parallel to $\overline{CD}$ in circle O.

1. If m∠*AOB* = 80 and m∠*COD* = 70, find m$\widehat{ADB}$.

2. If m$\widehat{BD}$ = 75 and m$\widehat{CD}$ = 90, find m∠*AOB*.

3. If m∠*BAO* = 40 and m$\widehat{CD}$ = 70, find m$\widehat{BD}$.

4. If *OC* = *CD* and m$\widehat{AC}$ = 79, find m∠*AOB*.

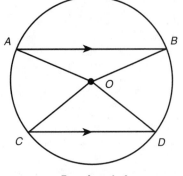

Exercises 1–4

5. If regular pentagon *ABCDE* is inscribed in circle *O*, what is the image of $R_{216°} \circ R_{-72°}(A)$?

6. In the accompanying diagram, line *p* is a line of symmetry of regular hexagon *ABCDEF*. Find the image of:
   (a) $R_{120°} \circ r_p(C)$
   (b) $r_p \circ R_{-240°}(B)$
   (c) $R_{120°} \circ r_p \circ R_{-60°}(A)$
   (d) $r_p \circ R_{-240°} \circ r_p(F)$

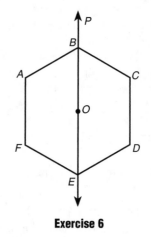

Exercise 6

7. Which of the following geometric figures has 45° rotational symmetry?
   (1) equilateral triangle   (3) rhombus
   (2) square   (4) regular octagon

8. A chord 48 centimeters in length is 7 centimeters from the center of a circle. What is the length of a radius of this circle?

9. A chord is 5 inches from the center of a circle whose diameter is 26 inches. What is the length of the chord?

10. How far from the center of a circle whose radius is 17 inches is a 30-inch chord?

11. In the accompanying figure, $\overline{JK}$, $\overline{JL}$, and $\overline{LK}$ are tangent to circle *O*. What is the length, *x*, of $\overline{JTL}$?

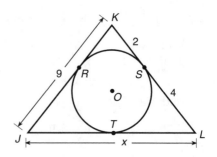

Exercise 11

6.2 MEASURING ANGLES OF CIRCLES

The vertex of an angle whose sides are chords, secants, or tangents may lie on the circle, inside the circle, or outside the circle. In each case there is a relationship between the measure of the angle and the measure(s) of the arc(s) it intercepts.

Vertex on the Circle: Inscribed Angles

In Figure 6.12, $\angle ABC$ is an *inscribed* angle that intercepts arc AC. An **inscribed angle** is an angle whose vertex is on the circle and whose sides are chords. If the measure of $\widehat{AC}$ is 50°, for example, then the measure of $\angle ABC$ is one-half of 50° or **25°**. Thus

$$m\angle ABC = \frac{1}{2}m\widehat{AC}.$$

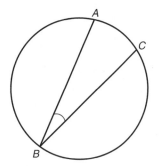

Figure 6.12 Inscribed Angles

INSCRIBED ANGLE THEOREM

The measure of an inscribed angle is one-half the measure of its intercepted arc.

As a result of the inscribed angle relationship:

- If two inscribed angles intercept the same (or congruent) arcs, then the angles must have the same measure and are, therefore, congruent. In Figure 6.13, inscribed angles B and D are congruent since they intercept the same arc, $\widehat{AC}$.

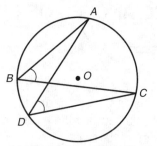

Figure 6.13 Angles That Intercept the Same Arcs. $\angle B \cong \angle D$

- An angle inscribed in a semicircle is a right angle since its intercepted arc is also a semicircle and $\frac{1}{2}(180°)$ is 90°. In Figure 6.14, $\angle AHB$ is a right angle since it is inscribed in semicircle AHB.

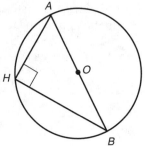

Figure 6.14 Angle Inscribed in a Semicircle

- The opposite angles of an inscribed quadrilateral are supplementary. In Figure 6.15,

$$a + c = 180$$

and

$$b + d = 180.$$

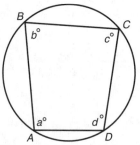

Figure 6.15 Opposite Angles of an Inscribed Quadrilateral

Examples

1. In each case, find the value of x:

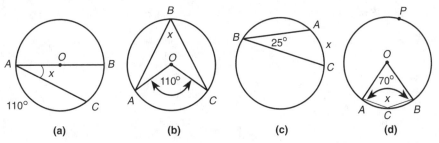

| (a) | (b) | (c) | (d) |

Solutions: **(a)** $\overset{\frown}{ABC}$ is a semicircle, so $m\overset{\frown}{BC} = 180 - 110 = 70$.

$$x = \frac{1}{2}m\overset{\frown}{BC} = \frac{1}{2}(70)$$
$$= \mathbf{35}$$

(b) $\angle AOC$ is a central angle, so $m\overset{\frown}{AC} = 110$.

$$m\angle ABC = \frac{1}{2}m\overset{\frown}{AC} = \frac{1}{2}(110)$$
$$= \mathbf{55}$$

(c) The measure of an arc intercepted by an inscribed angle must be twice the measure of the inscribed angle. Hence

$$x = 2(m\angle ABC) = 2(25)$$
$$= \mathbf{50}$$

(d) $\angle AOB$ is a central angle, so $m\overset{\frown}{ACB} = 70$.

$$m\overset{\frown}{APB} = 360 - 70 = 290$$
$$x = \frac{1}{2}m\overset{\frown}{APB} = \frac{1}{2}(290)$$
$$= \mathbf{145}$$

2. Given: In circle $O, \overline{AB}$ is a diameter, $\overset{\frown}{AM} \cong \overset{\frown}{DM}$.
   Prove: $\triangle AMB \cong \triangle CMB$

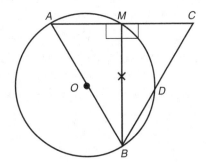

Solution: The triangles can be proved congruent by using the ASA postulate. Here is the formal proof.

PROOF

Statement	Reason
1. In circle O, $\overline{AB}$ is a diameter.	1. Given.
2. $\angle AMB$ is a right angle.	2. An angle inscribed in a semicircle is a right angle.
3. $\angle CMB$ is a right angle.	3. The supplement of a right angle is a right angle.
4. $\angle AMB \cong \angle CMB$ (A)	4. All right angles are congruent.
5. $\overline{MB} \cong \overline{MB}$ (S)	5. Reflexive property.
6. $\overparen{AM} \cong \overparen{DM}$	6. Given.
7. $\angle ABM \cong \angle CBM$ (A)	7. Inscribed angles that intercept congruent arcs are congruent.
8. $\triangle AMB \cong \triangle CMB$	8. ASA postulate.

Vertex on the Circle: Chord-Tangent Angle

In Figure 6.16, $\angle ABC$ is formed by chord $\overline{AB}$ and tangent $\overrightarrow{BC}$ and intercepts arc AB. If the measure of $\overparen{AB}$ is $110°$, for example, then the measure of $\angle ABC$ is one-half of $110°$ or $55°$. Thus

$$m\angle ABC = \frac{1}{2}\, m\overparen{AB}.$$

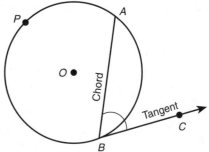

Figure 6.16 Chord-Tangent Angle

MATH FACTS

CHORD-TANGENT ANGLE THEOREM

The measure of an angle formed by a tangent and a chord drawn to the point of tangency is one-half the measure of its intercepted arc.

Example

3. In the accompanying diagram, $\overrightarrow{DE}$ is tangent to circle O, at D, $\overline{AOB}$ is a diameter, and $\overline{CD}$ is parallel to $\overline{AOB}$. If $m\angle DAB = 21$, find $m\angle CDE$.

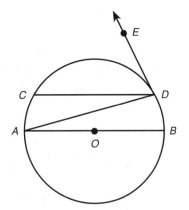

Solution: Angle CDE intercepts arc CD, so the measure of this arc is needed. Since the sum of the degree measures of the arcs of a semicircle is 180,

$$m\widehat{AC} + m\widehat{CD} + m\widehat{DB} = 180.$$

The degree measure of inscribed angle DAB is 21. The degree measure of its intercepted arc, $\widehat{DB}$, must be twice as great, or 42. Since parallel chords intercept equal arcs, $m\widehat{AC} = m\widehat{DB} = 42$. Hence:

$$42 + m\widehat{CD} + 42 = 180$$
$$m\widehat{CD} = 180 - 84$$
$$= 96$$

Since $\angle CDE$ is formed by a chord and a tangent,

$$m\angle CDE = \tfrac{1}{2}m\widehat{CD}$$
$$= \tfrac{1}{2}(96)$$
$$= \mathbf{48}$$

Vertex Inside the Circle: Angles Formed by Intersecting Chords

In Figure 6.17, $\angle AEC$ is formed by chords $\overline{AB}$ and $\overline{CD}$ intersecting at point E inside the circle. Angle AEC intercepts *two* arcs: arc AC and arc BD, which lies opposite its equal vertical angle. The measure of $\angle AEC$ is found by taking the average of the measures of its two intercepted arcs. Thus

$$m\angle AEC = \tfrac{1}{2}(m\widehat{AC} + m\widehat{BD}).$$

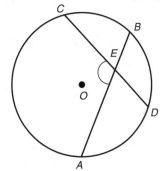

Figure 6.17 Angle Formed by Intersecting Chords

================ **MATH FACTS** ================

CHORD-CHORD ANGLE THEOREM

The measure of the angle formed by two chords intersecting inside a circle is one-half of the sum of the measures of its two intercepted arcs.

Examples

4. In the accompanying diagram, find the value of *x*.

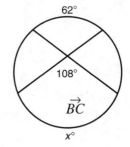

Solution: $108 = \frac{1}{2}(x + 62)$

or $2(108) = x + 62$

$216 = x + 62$

$x = 216 - 62 = \textbf{154}$

5. In the accompanying diagram, regular pentagon *ABCDE* is inscribed in circle *O*. Chords $\overleftrightarrow{AD}$ and $\overline{BE}$ intersect at *F*, and $\overleftrightarrow{BT}$ is tangent to circle *O* at *B*. Find:
(a) m∠*ABT*   **(b)** m∠*AFE*

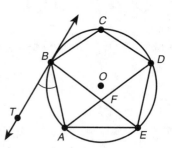

Solutions: **(a)** In a regular pentagon all of the five sides have the same length, and therefore circle *O* is divided into five equal arcs. The degree measure of each arc is, therefore, $\frac{360}{5}$ or 72. Hence

$$\text{m}\angle ABT = \tfrac{1}{2}\,\text{m}\widehat{AB} = \tfrac{1}{2}(72) = \textbf{36}.$$

(b) Angle *AFE* intercepts arcs *BCD* and *AE*. Since

$$\text{m}\widehat{AE} = 72 \text{ and } \text{m}\widehat{BCD} = \text{m}\widehat{BC} + \text{m}\widehat{CD} = 72 + 72 = 144,$$

$$\text{m}\angle AFE = \tfrac{1}{2}(\text{m}\widehat{BCD} + \text{m}\widehat{AE})$$

$$= \tfrac{1}{2}(144 + 72)$$

$$= \tfrac{1}{2}(216)$$

$$= \textbf{108}$$

6. In the accompanying diagram, chord $\overline{BE}$ bisects $\angle ABC$. If $m\angle ABC = 70$ and $m\widehat{BAE} = 200$, find $m\angle AFE$.

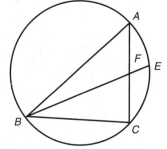

Solution: Angle *AFE* intercepts arcs *AE* and *BC*, whose measures must be determined. Since $\overline{BE}$ bisects inscribed angle *ABC*, $m\angle ABE = 35$, so $m\widehat{AE} = 2 \times 35$ or 70. Similarly, $m\angle CBE = 35$, so $m\widehat{CE} = 70$. Since the sum of the degree measures of the arcs of a circle is 360,

$$m\widehat{BC} + m\widehat{BAE} + m\widehat{CE} = 360$$
$$m\widehat{BC} + 200 \quad + 70 \quad = 360$$
$$m\widehat{BC} \quad = 90$$

The measure of $\angle AFE$ is one-half the sum of the measures of its intercepted arcs. Hence

$$m\angle AFE = \tfrac{1}{2}(m\widehat{AE} + m\widehat{BC})$$
$$= \tfrac{1}{2}(70 \quad + 90)$$
$$= \mathbf{80}$$

Vertex Outside the Circle: Angles Formed by Two Secants, Two Tangents, or a Secant and a Tangent

If the vertex of an angle lies outside the circle, the sides of the angle may be two secants, two tangents, or a secant and a tangent. In each case the measure of the angle is one-half the difference of the measures of the two arcs that lie between the sides of the angle (see Figure 6.18).

(1) Secant-Secant Angle	(2) Tangent-Tangent Angle	(3) Tangent-Secant Angle
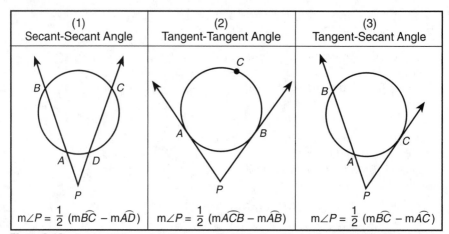		
$m\angle P = \frac{1}{2}\,(m\overset{\frown}{BC} - m\overset{\frown}{AD})$	$m\angle P = \frac{1}{2}\,(m\overset{\frown}{ACB} - m\overset{\frown}{AB})$	$m\angle P = \frac{1}{2}\,(m\overset{\frown}{BC} - m\overset{\frown}{AC})$

Figure 6.18 Angles Whose Vertices are Outside the Circle. In each case the measure of the angle is found by taking one-half the difference of the measures of the intercepted arcs.

MATH FACTS

SEC-SEC, TAN-TAN, TAN-SEC THEOREM

The measure of an angle formed by two secants, two tangents, or a tangent and a secant intersecting outside a circle is one-half the difference of the measures of its two intercepted arcs.

Examples

7. In each case, find the value of x:

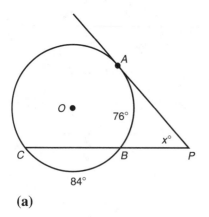

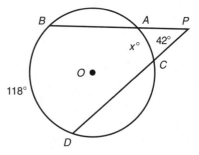

(a)   **(b)**

140

Solutions:

(a) $\text{m}\widehat{AC} + 76 + 84 = 360$

$\qquad \text{m}\widehat{AC} = 360 - 160$

$\qquad\qquad = 200$

Hence

$$x = \frac{1}{2}\,(\text{m}\widehat{AC} - \text{m}\widehat{AB})$$

$$= \frac{1}{2}\,(200 \ - 76)$$

$$= \frac{1}{2}\,(124)$$

$$= \mathbf{62}$$

(b) $\text{m}\angle P = \frac{1}{2}\,(\text{m}\widehat{BD} - \text{m}\widehat{AC})$

$\qquad 42 = \frac{1}{2}\,(118 \ - x)$

$\qquad 84 = 118 \qquad - x$

$\qquad\quad x = 118 - 84$

$\qquad\qquad = \mathbf{34}$

8. In the accompanying diagram, $\overrightarrow{PA}$ is tangent to circle O at point A. Secant $\overline{PBC}$ is drawn. Chords $\overline{CA}$ and $\overline{BD}$ intersect at point E. If
$$\text{m}\widehat{AD} : \text{m}\widehat{AB} : \text{m}\widehat{DC} : \text{m}\widehat{BC} = 2 : 3 : 4 : 6,$$
find the measure of each of the numbered angles.

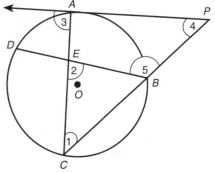

Solution: First find the degree measures of the arcs of the circle. Let $2x = \text{m}\widehat{AD}$. Then

$$3x = \text{m}\widehat{AB}, \quad 4x = \text{m}\widehat{CD}, \quad \text{and} \quad 6x = \text{m}\widehat{BC}.$$

Since the sum of the degree measures of the arcs of a circle is 360,

$$\text{m}\widehat{AD} + \text{m}\widehat{AB} + \text{m}\widehat{DC} + \text{m}\widehat{BC} = 360$$
$$2x \quad + 3x \quad + 4x \quad + 6x \quad = 360$$
$$15x = 360$$
$$x = \frac{360}{15} = 24$$

Hence

$$\text{m}\widehat{AD} = 2x = 2(24) = 48$$
$$\text{m}\widehat{AB} = 3x = 3(24) = 72$$
$$\text{m}\widehat{DC} = 4x = 4(24) = 96$$
$$\text{m}\widehat{BC} = 6x = 6(24) = 144$$

- $\angle 1$ is an inscribed angle:

    $m\angle 1 = \frac{1}{2} m\widehat{AB}$

    $\qquad = \frac{1}{2} (72)$

    $\qquad = 36$

- $\angle 3$ is a tangent-chord angle:

    $m\angle 3 = \frac{1}{2} m\widehat{AC} = \frac{1}{2} (m\widehat{AD} + m\widehat{DC})$

    $\qquad = \frac{1}{2} (48 \quad + 96)$

    $\qquad = 72$

- $\angle 2$ is a chord-chord angle:

    $m\angle 2 = \frac{1}{2} (m\widehat{AD} + m\widehat{BC})$

    $\qquad = \frac{1}{2} (48 \quad + 144)$

    $\qquad = 96$

- $\angle 4$ is a secant-tangent angle:

    $m\angle 4 = \frac{1}{2} (m\widehat{AC} - m\widehat{AB})$

    $\qquad = \frac{1}{2} (144 \quad - 72)$

    $\qquad = 36$

- $\angle 5$: Angles 5 and CBD are supplementary. The degree measure of $\angle 5$ may be found indirectly by first finding $m\angle CBD$.

$$m\angle CBD = \frac{1}{2} m\widehat{DC} = \frac{1}{2} (96) = 48$$

$$m\angle 5 = 180 - m\angle CBD$$
$$= 180 - 48$$
$$= 132$$

Exercise Set 6.2

1–8. Find the value of x.

1.

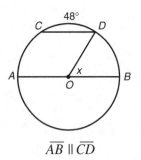

$\overline{AB} \parallel \overline{CD}$

2.

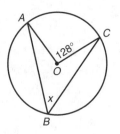

3.

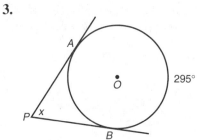

4.

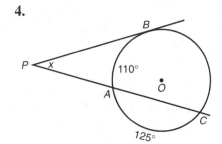

5.

 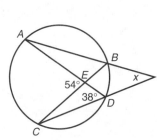

6.

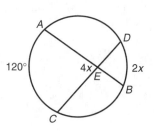

7.

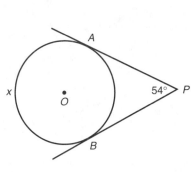

8.

9. In the accompanying diagram, $\overleftrightarrow{CD}$ is tangent to circle O at C. $\overline{AOB}$ is a diameter, and $\overline{OC}$ is a radius. Chords $\overline{AC}$ and $\overline{BC}$ are drawn. Find the degree measure of each of the numbered angles.

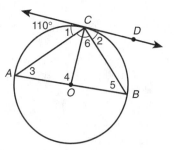

Exercise 9

143

10. Regular hexagon *ABCDEF* is inscribed in circle *O*. Chords $\overline{AD}$ and $\overline{CE}$ intersect at *H*. Tangents drawn to circle *O* at *A* and *F* intersect at point *P*. Find:
    (a) m∠*CED* **(b)** m∠*BAF* **(c)** m∠*AHE* **(d)** m∠*APF*

11. In the accompanying diagram, $\overline{CD}$ is tangent to circle *O*, $\overline{AB} \cong \overline{AD}$, m$\widehat{AB}$ = 150, and $\overline{EOD}$ is a diameter. Find:
    (a) m$\widehat{BD}$ **(c)** m∠*EFB*
    (b) m∠*ACD* **(d)** m∠*EDA*

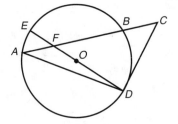

Exercise 11

12. In the accompanying diagram, $\overrightarrow{PCD}$ and $\overrightarrow{PBA}$ are secants from external point *P* to circle *O*. Chords $\overline{DA}$, $\overline{DEB}$, $\overline{CEA}$, and $\overline{CB}$ are drawn. $\overline{AB} \cong \overline{DC}$, m$\widehat{BC}$ is twice m$\widehat{AB}$, and m$\widehat{AD}$ is 60 more than m$\widehat{BC}$. Find:
    (a) m$\widehat{AB}$
    (b) m∠*P*
    (c) m∠*DAC*
    (d) m∠*DEA*
    (e) m∠*PCB*

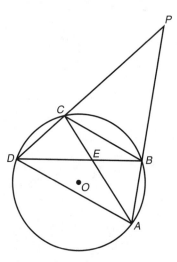

Exercise 12

13. In circle *O*, tangent $\overrightarrow{PW}$ and secant $\overrightarrow{PST}$ are drawn. Chord $\overline{WA}$ is parallel to chord $\overline{ST}$. Chords $\overline{AS}$ and $\overline{WT}$ intersect at point *B*. If m$\widehat{WA}$: m$\widehat{AT}$: m$\widehat{ST}$ = 1 : 3 : 5, find:
    (a) m$\widehat{WA}$, m$\widehat{AT}$, m$\widehat{ST}$, m$\widehat{SW}$
    (b) m∠*WTS*
    (c) m∠*TBS*
    (d) m∠*TWP*
    (e) m∠*WPT*
    (f) m∠*ASP*

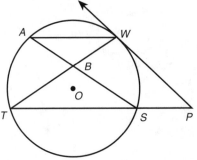

6.3 FINDING CHORD, TANGENT, AND SECANT-SEGMENT LENGTHS

If two angles of one triangle are congruent to two angles of another triangle, then the triangles are similar and the lengths of their corresponding sides are in proportion. If triangles that have chord, tangent, or secant segments as sides can be proved similar, some useful proportions involving the lengths of these segments can be formed.

Similar Triangles and Circles

The following angle measurement relationships are often used in obtaining the congruent pairs of angles needed to prove that triangles are similar:

- Inscribed angles (or angles formed by a tangent and a chord) that intercept the same (or congruent) arcs are congruent.
- Angles that are inscribed in a semicircle, are formed by a radius drawn to a point of tangency, or are formed by perpendicular lines are right angles and are, therefore, congruent.

Example

1. Given: In circle P, $\overline{AB}$ is a diameter, $\overline{DB}$ is a tangent to circle P at B, $\overline{CD} \perp \overline{DB}$.

Prove: $\dfrac{AB}{BC} = \dfrac{BC}{CD}$

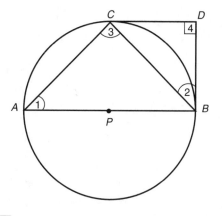

Solution: To prove that a proportion is true, first identify the triangles that have these segments as sides, and then prove that the triangles are similar. In this example, it is necessary to prove that triangles ABC and CBD are similar since $\overline{AB}$ and $\overline{BC}$ (reading "across" the *top* of the proportion) are sides of $\triangle ABC$, and $\overline{BC}$ and $\overline{CD}$ (reading "across" the *bottom* of the proportion) are sides of $\triangle CBD$.

Proof: Angle 1 is an inscribed angle, and $\angle 2$ is formed by a tangent and a chord. Since the measure of each angle is one-half the measure of the same arc, $\overset{\frown}{BC}$, angles 1 and 2 are congruent. Angle 3 is inscribed in a semicircle, and $\angle 4$ is formed by perpendicular lines. Since each angle is a right angle, angles 3 and 4 are congruent. Hence

$$\angle 1 \cong \angle 2 \quad \text{and} \quad \angle 3 \cong \angle 4,$$

so triangles ABC and CBD are similar by the AA Theorem of Similarity. The lengths of corresponding sides of similar triangles are in proportion, so

$$\frac{AB}{BC} = \frac{BC}{CD}.$$

Products of Segments of Intersecting Chords

In Figure 6.19, chords $\overline{AB}$ and $\overline{CD}$ intersect at E, forming segments p, q, r, and s as shown. Triangles AED and CEB are similar since (1) the vertical angles at E are congruent, and (2) inscribed angles D and B are congruent because they intercept the same arc.

Since the lengths of corresponding sides of similar triangles are in proportion,

$$\frac{p}{r} = \frac{s}{q} \quad \text{or} \quad p \cdot q = r \cdot s.$$

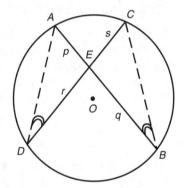

Figure 6.19 Products of Intersecting-Chord Segments

MATH FACTS

CHORD SEGMENTS THEOREM

If two chords intersect, the product of the segment lengths of one chord is equal to the product of the segment lengths of the other chord.

Examples

2. In the accompanying diagram, chords $\overline{AB}$ and $\overline{CD}$ of circle O intersect in E so that $CE = 5$ centimeters and $ED = 16$ centimeters. If the length of $\overline{AE}$ exceeds the length of $\overline{EB}$ by 2 centimeters, find the length of $\overline{EB}$.

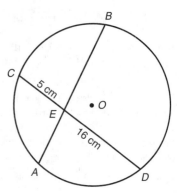

Solution: Let $EB = x$. Then $AE = x + 2$.

$$EB \cdot AE = CE \cdot ED$$
$$x(x+2) = \ 5 \cdot 16$$
$$x^2 + 2x - 80 = 0$$
$$(x + 10)(x - 8) = 0$$

$x + 10 = 0 \qquad$ or $\quad x - 8 = 0$

$\qquad x = -10 \qquad\qquad\quad x = 8$

Reject -10 since x must be positive.

$EB = x = 8$ and $AE = x + 2 = 8 + 2 = 10$

The length of $\overline{EB}$ is **8 cm**.

3. In circle O, chord $\overline{AB}$ bisects chord $\overline{DC}$ at E. If $AE = 4$ and $EB = 89$, find the length of $\overline{DC}$.

Solution: Since $\overline{AB}$ bisects $\overline{DC}$ at E, point E is the midpoint of $\overline{DC}$. Let $x = CE = ED$.

$CE \cdot ED = EB \cdot AE$

$x \cdot \quad x = \quad 9 \cdot 4$

$\qquad x^2 = 36$

$\qquad x = \ \sqrt{36} \quad = 6$

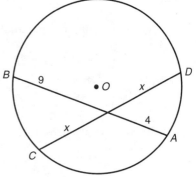

Hence $CE = ED = 6$. Since $CD = CE + ED = 12$, the length of $\overline{CD}$ is **12**.

Tangent-Secant Segments

In Figure 6.20, tangent $\overline{PT}$ and secant $\overline{PAB}$ are drawn from the same point, P. Secant segment $\overline{PA}$ falls outside the circle and is called an *external secant segment*. By drawing $\overline{AT}$ and $\overline{BT}$, triangles PAT and PTB are formed. These triangles are similar since (1) $\angle P \cong \angle P$, and (2) the measures of angles PTA and B are each one-half the measure of $\overline{AT}$, and the angles are, therefore, congruent. Hence

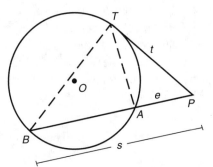

$$\frac{t}{e} = \frac{s}{t} \quad \text{and} \quad t^2 = s \cdot e.$$

Figure 6.20 Tangent-Secant Segments

$$\boxed{\text{Math Facts}}$$

Tan-Sec Segments Theorem

If a tangent and a secant are drawn to a circle from the same point, then the square of the length of the tangent equals the product of the lengths of the secant and its external segment.

Example

4. In the accompanying diagram, $\overline{PT}$ is tangent to circle O at T and $\overline{PAB}$ is a secant.
(a) If $PT = 6$ and $PAB = 12$, find PA.
(b) If $PA = 4$ and $AB = 12$, find PT.
(c) If $PT = 12$ and $AB = 7$, find PAB.

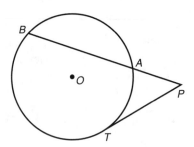

Solutions: **(a)** $(PT)^2 = PAB \cdot PA$
$$6^2 = 12(PA)$$
$$36 = 12(PA)$$
$$\text{or} \quad PA = \frac{36}{12}$$
$$= 3$$

(b) Since $PA = 4$ and $AB = 12$, $PAB = 4 + 12 = 16$.

$$(PT)^2 = PAB \cdot PA$$
$$= (16)(4)$$
$$= 64$$
$$PT = \sqrt{64}$$
$$= \mathbf{8}$$

(c) Let $PAB = x$. Then $PA = x - 7$.

$$(PT)^2 = PAB \cdot PA$$
$$(12)^2 = x(x - 7)$$
$$144 = x^2 - 7x$$
$$0 = x^2 - 7x - 144 \quad \text{or} \quad x^2 - 7x - 144 = 0$$
$$(x - 16)(x + 9) = 0$$

$$x - 16 = 0 \qquad \text{or} \qquad x + 9 = 0$$
$$x = 16 \qquad\qquad\quad x = -9. \text{ Reject since length}$$
$$\text{cannot be negative.}$$

$PAB = \mathbf{16}$

Segment of Two Secants

If in Figure 6.21 two secants are drawn to a circle from the same point, then

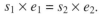

$$s_1 \times e_1 = s_2 \times e_2.$$

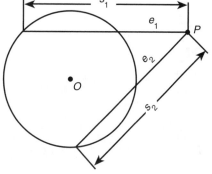

Figure 6.21 Segments of Two Secants

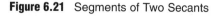

Unit Two **FUNCTIONS AND CIRCLES**

Example

5. In each case find the value of *x*:

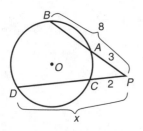

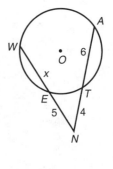

(a)   **(b)**

Solutions:

(a) $PA \cdot PB = PC \cdot PD$
$$3 \cdot 8 \ = 2 \ \cdot x$$
$$24 = 2x$$
$$\mathbf{12} = x$$

(b) $NE \cdot NW = NT \cdot NA$, where $NE = 5$,
$NW = x + 5$, $NT = 4$, $NA = 6 + 4 = 10$
$$5(x + 5) = 4 \cdot 10$$
$$5x + 25 = 40$$
$$5x = 15$$
$$x = \mathbf{3}$$

Exercise Set 6.3

1. If $RK = 5$, $KH = 9$, and $PK = 15$, find KN.

2. If $PK = 27$, $KN = 3$, and K is the midpoint of $\overline{RH}$, find RK.

3. If $RH = 16$, $RK = 4$, and $PK = 8$, find PN.

4. If $RH = 22$, $PK = 7$, and $KN = 3$, find RK.

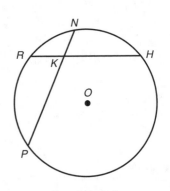

Exercises 1–4

5. If $PW = 5$, $PG = 8$, and $PH = 2$, find KH.

6. If $GW = 7$, $PW = 3$, and $PK = 15$, find PH.

7. If W is the midpoint of $\overline{GP}$ and $PH = 5$, and $KH = 35$, find PG.

8. If $PW = 6$, $WG = 9$, and $PH = 9$, find KH.

9. If $GW = 11$, $PH = 5$, and $KH = 7$, find PW.

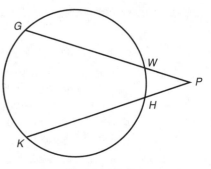

Exercises 5–9

10. If $RV = 9$ and $RM = 3$, find RT.

11. If $MT = 24$ and $RM = 1$, find RV.

12. If $RV = 8$ and $RM = 4$, find MT.

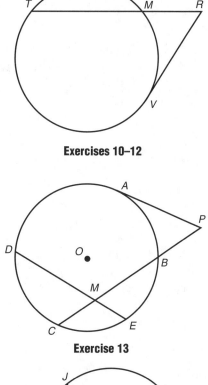

Exercises 10–12

13. Given: circle O with tangent $\overline{PA}$ and secant $\overline{PBC}$; chord $\overline{DE}$ bisects chord $\overline{BC}$ at M; $PA = 3$; $PB = 1$.
  (a) Find BC.
  (b) If $DE = 10$ and $DM > EM$, find EM.

Exercise 13

14. (a) If $NA = 4$, $JA = 8$, and $WA = 11$, find OK.
  (b) If A is the midpoint of $\overline{JW}$, $AK = 32$, and $NA = 18$, find JW and OA.
  (c) If $JA = 12$, $AW = 9$, and AK is three times the length of $\overline{NA}$, find OK.

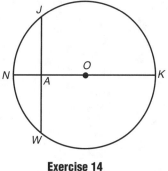

Exercise 14

15. A circle divides a secant segment into an internal segment having a length of 30 and an external segment having a length of 2. Find the length of a tangent segment drawn to the circle from the same exterior point.

16. From a point 2 units from a circle, a tangent segment is drawn. If the radius of the circle is 8, find the length of the tangent segment.

17. Secant $\overleftrightarrow{PAB}$ and tangent $\overline{PC}$ are drawn to circle O from point P. If the ratio of PA to AB is 1 to 4 and the length of PC is 80, find the length of secant $\overline{PAB}$.

18. Given: Triangle HBW is inscribed in circle O, tangent segment $\overline{AB}$ is tangent at point B, $ABLM$ is a parallelogram.

    Prove: $\dfrac{BL}{BW} = \dfrac{BM}{BH}$

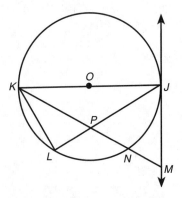

Exercise 18

19. Given: In circle O, $\overline{KJ}$ is a diameter, $\overleftrightarrow{MJ}$ is tangent at point J, N is the midpoint of $\overarc{LNJ}$.

    Prove: $\dfrac{KL}{KJ} = \dfrac{KP}{KM}$

20. Given: In circle O, $\overline{KJ}$ is a diameter, $\overleftrightarrow{MJ}$ is tangent at point J, $\overline{JP} \cong \overline{JM}$.

    Prove: $\dfrac{KL}{KJ} = \dfrac{LP}{KM}$

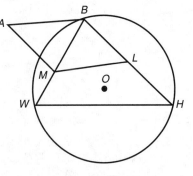

Exercises 19 and 20

Each of the questions in this section has appeared on a previous Course III Regents Examination. Here is an opportunity for you to review Chapter 6 and, at the same time, prepare for the Course III Regents Examination.

1. In the accompanying diagram, $\overrightarrow{PA}$ is tangent to circle O at A and $\overline{PBC}$ is a secant. If $CB = 9$ and $PB = 3$, find the length of $\overline{PA}$.

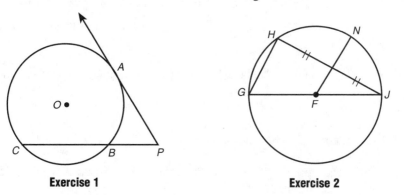

Exercise 1   **Exercise 2**

2. In the accompanying diagram of circle F, $\overline{FN}$ bisects $\overline{HJ}$ and m$\widehat{NJ} = 68$. Find m$\widehat{HG}$.

3. In circle O, diameter $\overline{AB}$ is perpendicular to chord $\overline{CD}$ at E. If $AE = 16$ and $EB = 4$, find CD.

4. Two secants, $\overline{ABC}$ and $\overline{ADE}$, are drawn to a circle from external point A. If $AB = 4$, $BC = 6$, and $AD = 5$, find DE.

5. In the accompanying diagram of circle O, chords $\overline{AB}$ and $\overline{CD}$ intersect at E and $\overline{AD}$ is a diameter. If m$\widehat{CB} = 82$, find m$\angle AED$.

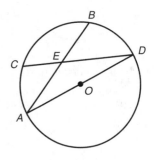

6. In the accompanying diagram of circle O, diameter $\overline{AB}$ is perpendicular to chord $\overline{CD}$ at point E. What is the image of $\overline{AC}$ in $\overline{AB}$?

(1) $\overline{AD}$ (3) $\overline{ED}$

(2) $\overline{BD}$ (4) $\overline{AE}$

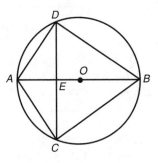

7. Quadrilateral $ABCD$ is inscribed in circle O. If $m\widehat{AB} = 132$ and $m\widehat{BC} = 82$, find $m\angle ADC$.

8. In circle O, chord $\overline{CD}$ is parallel to diameter $\overline{AB}$ with radii $\overline{OC}$ and $\overline{OD}$ drawn. If $m\widehat{AC} = 25$, what is $m\angle COD$?

9. In a circle, chords $\overline{AB}$ and $\overline{CD}$ intersect at point E. If $AE = x + 1$, $EB = x$, $CE = 2$, and $ED = 3$, find the value of x.

10. In the accompanying diagram, $\overrightarrow{PC}$ is tangent to circle O at point C. If $m\widehat{AD} = 122$ and $m\angle BAC = 73$, find:

(a) $m\widehat{BC}$ **(b)** $m\angle ABC$ **(c)** $m\angle P$ **(d)** $m\angle BEA$ **(e)** $m\angle PDA$

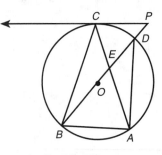

Exercise 10

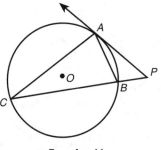

Exercise 11

11. In the accompanying diagram, $\overrightarrow{PA}$ is tangent to circle O at A, $\overline{PBC}$ is a secant, $m\angle P = 42$, and $m\widehat{AC}:m\widehat{AB} = 5:2$. Find:

(a) $m\widehat{AC}$ **(b)** $m\widehat{BC}$ **(c)** $m\angle ACB$ **(d)** $m\angle ABP$

12. In the accompanying diagram of circle O, m$\widehat{AB}$: m$\widehat{BC}$ = 1 : 2; diameter $\overline{CA}$ and chord $\overline{AE}$ are drawn; chord $\overline{EC}$ is parallel to chord $\overline{AB}$; chord $\overline{BC}$ is extended through C to D, and tangent $\overline{DE}$ is drawn. Find:

   (a) m$\widehat{BC}$

   (b) m$\widehat{CE}$

   (c) m$\angle AEC$

   (d) m$\angle CED$

   (e) m$\angle BDE$

ANSWERS TO SELECTED EXERCISES: CHAPTER 6

Section 6.1

1. 210   **5.** C **9.** 24 in.

3. 95   **7.** (4)   **11.** 11

Section 6.2

1. 66°   **9.** m$\angle 1$ = 55, m$\angle 2$ = 35, m$\angle 3$ = 35,

3. 115°   m$\angle 4$ = 110, m$\angle 5$ = 55, m$\angle 6$ = 55

5. 232°   **11. (a)** 60   **(b)** 45   **(c)** 135   **(d)** 15

7. 22°   **13. (a)** m$\widehat{WA}$ = 30, m$\widehat{AT}$ = 90, m$\widehat{ST}$ = 150, m$\widehat{SW}$ = 90

             (b) 45   **(c)** 90   **(d)** 120   **(e)** 15   **(f)** 135

Section 6.3

1. 3   **5.** 18   **9.** 4   **13. (a)** 8   **15.** 8

3. 6   **7.** 20   **11.** 5   **(b)** 2   **17.** 20

18. Prove $\triangle BLM \sim \triangle BWH$ by using the AA Theorem of Similarity:

- $\angle MBL \cong HBW$ (Reflexive property)
- $\angle ABM \cong \angle BML$ (Alternate interior angles formed by parallel lines are congruent.)
- $\triangle BLM \sim \triangle BWH$, so $\frac{BL}{BW} = \frac{BM}{BH}$ (The lengths of corresponding sides of similar triangles are in proportion.)

19. Prove $\triangle KLP \sim \triangle KJM$ by using the AA Theorem of Similarity:

- $\angle L \cong \angle KJM$ (All right angles are congruent.)
- m$\angle LKP = \frac{1}{2}$ m$\widehat{NC}$, m$\angle JKM = \frac{1}{2}$ m$\widehat{JN}$, and $\overline{JN} \cong \overline{NL}$, so $\angle LKP \cong \angle JKM$.
- $\triangle KLP \sim \triangle KJM$, so $\frac{KL}{KJ} = \frac{KP}{KM}$ (The lengths of corresponding sides of similar triangles are in proportion.)

20. Prove $\triangle KLP \sim \triangle KJM$ by using the AA Theorem of Similarity:

- $\angle L \cong \angle KJM$ (All right angles are congruent.)
- Since $\overline{JP} \cong \overline{JM}$, then $\angle M \cong \angle JPM \cong \angle KPL$.
- $\triangle KLP \sim \triangle KJM$, so $\frac{KL}{KJ} = \frac{LP}{KM}$ (The lengths of corresponding sides of similar triangles are in proportion.)

Regents Tune-Up: Chapter 6

1. 6	**6.** (1)	**10. (a)** 146	**11. (a)** 140	**12. (a)** 120
2. 44	**7.** 107	**(b)** 78	**(b)** 164	**(b)** 60
3. 16	**8.** 130	**(c)** 56	**(c)** 28	**(c)** 90
4. 3	**9.** 2	**(d)** 46	**(d)** 110	**(d)** 30
5. 131		**(e)** 151		**(e)** 60

Unit Three TRIGONOMETRY AND SOLVING TRIANGLES

Chapter 7

TRIGONOMETRIC FUNCTIONS: THE GENERAL ANGLE

7.1 REVIEWING RIGHT-TRIANGLE TRIGONOMETRY

∧
KEY IDEAS
∠ ∖

In a right triangle, the ratios of the lengths of selected pairs of sides can be related to the measure of either of the acute angles, using the sine, cosine, and tangent trigonometric ratios:

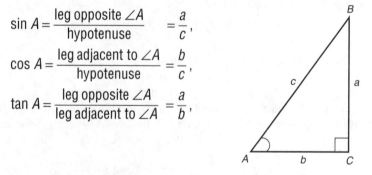

$$\sin A = \frac{\text{leg opposite } \angle A}{\text{hypotenuse}} = \frac{a}{c},$$

$$\cos A = \frac{\text{leg adjacent to } \angle A}{\text{hypotenuse}} = \frac{b}{c},$$

$$\tan A = \frac{\text{leg opposite } \angle A}{\text{leg adjacent to } \angle A} = \frac{a}{b},$$

where *sin*, *cos*, and *tan* are abbreviations for *sine*, *cosine*, and *tangent*, respectively.

Writing Trigonometric Ratios

If the lengths of two sides of a right triangle are known, the trigonometric function values of either acute angle can be determined.

Example

1. In right triangle *ACB*, hypotenuse *AB* = 5 and *BC* = 3. Find the value of:
(a) sin *A*   **(b)** cos *A*   **(c)** tan *A*

Solutions: Since $\triangle ACB$ is a *3-4-5* right triangle, $AC = 4$.

(a) $\sin A = \dfrac{\text{leg opposite } \angle A}{\text{hypotenuse}}$

$= \dfrac{BC}{AB} = \dfrac{3}{5}$ or **0.6**

(b) $\cos A = \dfrac{\text{leg adjacent to } \angle A}{\text{hypotenuse}}$

$= \dfrac{AC}{AB} = \dfrac{4}{5}$ or **0.8**

(c) $\tan A = \dfrac{\text{leg opposite } \angle A}{\text{leg adjacent to } \angle A}$

$= \dfrac{BC}{AC} = \dfrac{3}{4}$ or **0.75**

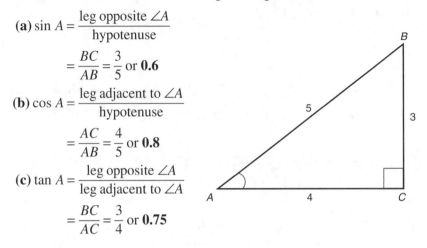

Finding Trigonometric Function Values

To use a scientific calculator to find the value of sin x, cos x, or tan x, where x is measured in degrees, proceed as follows:

- Set the angular mode of your calculator to DEGrees, if necessary. If you see RADians instead of DEGrees in the display window, press the $\boxed{\text{DRG}}$ or $\boxed{\text{MODE}}$ key until DEG appears in the display window.
- Press the key labeled with the name of the appropriate trigonometric function, enter the number x, and then press the $\boxed{=}$ key. Some calculators require that you enter the measure of the angle before pressing the trignometric function key. For example, to find the value of sin 54°, use the key sequence

$$\boxed{\text{SIN}} \quad 54 \quad \boxed{=} \qquad or \qquad 54 \quad \boxed{\text{SIN}}.$$

Thus, sin 54° ≈ 0.8090169.

Converting Minutes to Degrees

Each of the 60 equal parts of a degree is called a **minute**. The angle 28°30′ is read as "28 degrees, 30 minutes." Since there are 60 minutes in 1 degree, dividing 30 minutes by 60 changes 30 minutes to a fractional part of a degree. Thus

$$28°30' = 28° + \left(\frac{30}{60}\right)° = \mathbf{28.5°}.$$

CHANGING FROM MINUTES TO DEGREES

To convert $n°m'$ to an equivalent number of degrees, divide m by 60 and then add the result to n. Some scientific calculators have a function key that converts degrees and minutes into degrees and decimal parts of a degree.

Example

2. Find the value of cos 37°20′ correct to *four decimal places.*

Solution: If your calculator does not have a key that converts a minute into a decimal part of a degree, change 37°20′ into an equivalent number of degrees by dividing 20 by 60, adding the quotient to 37, and then evaluating the cosine of the resulting angle:

$$20 \boxed{÷} \ 60 \boxed{+} \ 37 \boxed{=} \ \boxed{\text{COS}}.$$

If this key sequence does not work with your calculator, try this sequence:

$$\boxed{\text{COS}} \ \boxed{(} \ 20 \boxed{÷} \ 60 \boxed{+} \ 37 \boxed{)} \ \boxed{=}.$$

Finding Angles Using Inverse Trigonometric Functions

Most scientific calculators have $\sin^{-1}$ printed directly above the $\boxed{\text{SIN}}$ key, $\cos^{-1}$ printed directly above the $\boxed{\text{COS}}$ key, and $\tan^{-1}$ printed directly above the $\boxed{\text{TAN}}$ key. These second or *inverse* calculator functions are used to find the measure of an angle when you know the value of a particular trigonometric function of that angle. For example, if $\tan x = 2.197$, then $x = \tan^{-1} 2.197$, which is read as "x is an angle whose tangent is 2.197."

To find the degree measure of $\angle x$, make sure the angular mode of the calculator is set to degrees. Then key in

$$2.197 \ \boxed{2 \text{ nd}} \ \boxed{\text{TAN}^{-1}} \quad \text{or} \quad \boxed{2 \text{ nd}} \ \boxed{\text{TAN}^{-1}} \ 2.197 \ \boxed{=}.$$

Since $x \approx 65.526579°$, $\angle x$, correct to the *nearest degree,* is **66**.

Note: One way to change 65.526579° to an angle expressed to the *nearest minute* is to multiply the first three digits of the decimal part by 60, round off the product, and then add the result to 65. Since $.526 \times 60' \approx 32'$, $65.526579° \approx 65°32'$. Some calculators have a special function key that does this operation automatically.

Angles of Elevation and Depression

A person sometimes has to look up or down to sight a distant object. The **angle of elevation** is the angle through which an observer must *raise* his or her line of sight in order to see an object *above* the observer's horizontal line of vision. The **angle of depression** is the angle through which an observer must *lower* his or her line of sight in order to see an object *below* the observer's horizontal line of vision. As shown in Figure 7.1, the measures of the angles of elevation and depression are equal, since they are alternate interior angles formed by parallel lines.

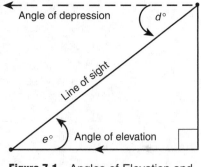

Figure 7.1 Angles of Elevation and Depression

Examples

3. A man standing 30 feet from a flagpole observes the angle of elevation of its top to be 48°30′. Find the height of the flagpole, correct to the *nearest tenth of a foot*.

Solution: If x = the height of the flagpole, then

$$\tan 48°30′ = \frac{x}{30}$$
$$x = 30 \tan 48°30′$$
$$= 30(1.1303)$$
$$= 33.909$$

The height of the flagpole, correct to the *nearest tenth of a foot,* is **33.9**.

4. When the altitude of a plane is 800 meters, the pilot spots a target at a distance of 1200 meters. Find, correct to the *nearest 10 minutes,* the angle of depression at which the pilot observes the target.

Solution: If x = measure of angle of depression, then

$$\sin x° = \frac{800}{1200} = \frac{2}{3} = 0.6667$$
$$x = \sin^{-1} 0.6667$$
$$\approx 41.813°$$
$$\approx 41°49′$$

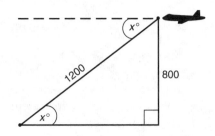

The angle of depression, correct to the *nearest 10 minutes,* is **41°50′**.

Exercise Set 7.1

1. Find to four decimal places:
   (a) $\sin 30°$ (c) $\cos 67°10′$ (e) $\tan 18°20′$
   (b) $\tan 45°$ (d) $\sin 55°40′$ (f) $(\sin 60°)^2 + (\cos 60°)^2$

2. Find the complement of each of the following angles:
   (a) $19°20′$ (b) $35°42′$ (c) $64°52′$

3. Find the supplement of each of the following angles:
   (a) $64°25′$ (b) $113°38′$ (c) $126°08′$

4. Find the degree measure of angle x, correct to the *nearest 10 minutes*:
   (a) $\cos x = 0.8700$ (c) $\tan x = 1.2110$ (e) $\tan x = 0.5036$

   (b) $\sin x = 0.9298$ (d) $\cos x = \dfrac{3}{5}$ (f) $\sin x = \dfrac{5}{8}$

5. In right triangle RTS, $\angle T$ is the right angle. If $\sin R = \dfrac{5}{13}$, express as a fraction:
   (a) $\cos R$ and $\tan R$ (b) $\sin S$, $\cos S$, and $\tan S$

6. If the length of a diagonal of a rectangle is 41 and the length of its longest side is 40, find, correct to the *nearest 10 minutes,* the degree measure of the angle formed by the diagonal and the
   (a) shortest side   (b) longest side

7. In an isosceles right triangle, the length of the hypotenuse is 2. Find:
   (a) the length of each leg
   (b) the values of the sine, cosine, and tangent of either acute angle of the triangle

8. The perimeter of an equilateral triangle is 6. If an altitude is drawn to the base of the triangle, find:
   (a) the length of the altitude
   (b) the values of the sine, cosine, and tangent of the angle formed by the base and a side of the triangle. (Answers may be left in radical form.)

9. In right triangle ACB, $\angle C$ is the right angle, $AB = 25$, and $AC = 24$. Find the values of the sine, cosine, and tangent of the acute angle having the *smaller* degree measure.

10. At noon, a tree having a height of 10 feet casts a shadow 14 feet in length. Find, to the *nearest tenth of a degree,* the angle of elevation of the sun at this time.

11. Find, to the *nearest tenth of a foot,* the height of a building that casts a shadow 80.4 feet when the angle of elevation of the sun is 42°40′.

12. A man observes that the angle of depression from the top of a cliff overlooking the ocean to a ship is 37°30′. If at this moment the ship is 500 meters from the foot of the cliff, find, to the *nearest tenth of a meter,* the height of the cliff.

13. When the altitude of a plane is 950 meters, the pilot spots a target at a distance of 1275 meters. At what angle of depression, correct to the *nearest 10 minutes,* does the pilot observe the target?

14. At Mogul's Ski Resort, the beginner's slope is inclined at an angle of 12.3°, while the advanced slope is inclined at an angle of 26.4°. If Rudy skis 1000 meters down the advanced slope while Valerie skis the same distance on the beginner's slope, how much longer was the horizontal distance that Valerie covered?

15. In the diagram shown, ACB is an isosceles right triangle, $m\angle C = 90$, $AC = 10$, and $\overline{AD}$ bisects $\angle BAC$. Find to the *nearest tenth* the length of:
   (a) $\overline{CD}$ (b) $\overline{BD}$

7.2 PLACING ANGLES IN STANDARD POSITION

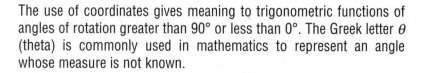
KEY IDEAS

The use of coordinates gives meaning to trigonometric functions of angles of rotation greater than 90° or less than 0°. The Greek letter θ (theta) is commonly used in mathematics to represent an angle whose measure is not known.

Standard Position

An angle is in **standard position** when its vertex is at the origin (O) and one of its sides, called the **initial side,** remains fixed on the positive x-axis (see Figure 7.2). The other side of the angle, called the **terminal** side, moves in a counterclockwise direction for positive angles of rotation.

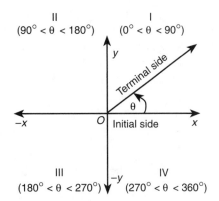

Figure 7.2   Standard Position of an Angle

Angles Greater Than 360°

An angle in standard position whose measure is greater than 360° will complete more than one full rotation. For example, an angle of rotation of 410° will complete one full rotation and its terminal side will lie in Quadrant I, making an angle of 410°–360° or 50° with the positive x-axis.

Although the angles 50° and 410° have different measures, they are *coterminal*. **Coterminal angles** (see Figure 7.3) are angles whose terminal sides coincide. To reduce an angle greater than 360° to a coterminal angle between 0° and 360°, successively subtract 360° from the given angle until the difference is between 0 and 360. For example, to find an angle that is coterminal with an angle of 860°, proceed as follows:

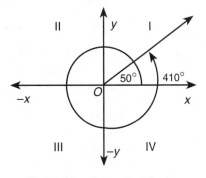

Figure 7.3   Coterminal Angles

Subtract 360 from 860: $860 - 360 = 500$
Subtract 360 from 500: $500 - 360 = \mathbf{140}$

Hence angles of 860° and 140° are coterminal.

Negative Angles

A *negative* angle indicates a clockwise rotation about the origin. For example, the terminal side of an angle measuring –150° lies in Quadrant III, as shown in Figure 7.4. Notice that angles of –150° and 210° are coterminal.

163

A positive and negative angle are coterminal if the sum of the absolute values of their measures is 360° (or a multiple of 360°).

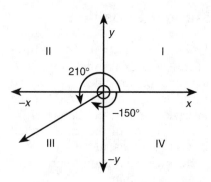

Figure 7.4 Negative Angles

Arc Length and Radian Measure

A *radian,* like a degree, is a unit of angle measure. The radian measure of a central angle of a circle compares the length of its intercepted arc with the length of the radius of the circle.

MATH FACTS

DEFINITION OF RADIAN AND ARC-LENGTH FORMULA

One **radian** is the measure of a central angle of a circle that intercepts (subtends) an arc whose length equals the radius of the circle.

If θ represents the measure of a central angle in radians and s represents the length of its intercepted arc, then, in a circle whose radius is r,

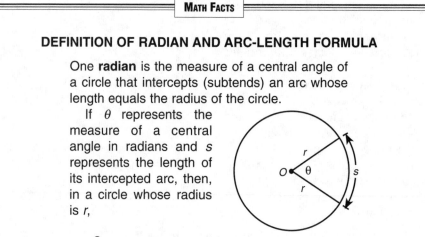

$$\theta = \frac{s}{r} \quad \text{or} \quad s = r\theta.$$

Example

If the measure of a central angle of a circle is 2 radians and the radius of the circle is 5 centimeters, then the length of the intercepted arc s is

$$s = r\theta = 5 \times 2 = 10 \text{ cm}.$$

Converting Between Radians and Degrees

Since angles may be measured in degrees or in radians, it is sometimes necessary to be able to express the degree measure of a given angle in radians, or the radian measure of a given angle in degrees.

- To change from degrees to radians:

  Multiple the number of degrees by $\dfrac{\pi}{180°}$.

  Example: $60° = \overset{1}{\cancel{60°}} \times \left(\dfrac{\pi}{\underset{3}{\cancel{180°}}} \right) = \dfrac{\pi}{3}$ **radians**

- To change from radians to degrees:

  Multiply the number of radians by $\dfrac{180°}{\pi}$.

  Example: $\dfrac{\pi}{2}$ radians $= \dfrac{\cancel{\pi}}{\cancel{2}} \times \dfrac{\overset{90}{\cancel{180°}}}{\cancel{\pi}} = \mathbf{90}°$

Examples

1. Find the measure of a central angle that intercepts an arc of length 3π in a circle whose radius is 9 centimeters.

Solution: In a circle, the length of an arc s equals the product of its radius r and the radian measure of its central angle θ. Since $s = r\theta$ and $s = 3\pi$, and $r = 9$:

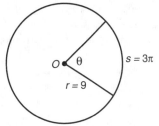

$3\pi = 9\theta \quad$ or $\quad \theta = \dfrac{3\pi}{9} = \dfrac{\pi}{3}$ **radians**

2. Change from degrees to radians:
(a) $45°$ **(b)** $300°$

Solutions: To convert from degrees to radians, multiply by $\dfrac{\pi}{180°}$.

In each case leave the answer in terms of π as this form gives an *exact* value.

(a) $\cancel{45°} \times \dfrac{\pi}{\underset{4}{\cancel{180}}} = \dfrac{\pi}{4}$ **radians**   **(b)** $\overset{5}{\cancel{300°}} \times \dfrac{\pi}{\underset{3}{\cancel{180°}}} = \dfrac{5\pi}{3}$ **radians**

3. Change from radians to degrees:

(a) $\dfrac{7}{12}\pi$ (b) $-\dfrac{5\pi}{9}$

Solutions: To convert from radians to degrees, multiply by $\dfrac{180°}{\pi}$.

(a) $\dfrac{7\overset{15°}{\cancel{\pi}}}{\cancel{12}} \times \dfrac{\cancel{180°}}{\cancel{\pi}} = \mathbf{105\,°}$ (b) $-\dfrac{5\overset{20°}{\cancel{\pi}}}{\cancel{9}} \times \dfrac{\cancel{180°}}{\cancel{\pi}} = \mathbf{-100\,°}$

Scientific Calculators and Radian Measure

To evaluate trigonometric functions of angles expressed in radians, first set the angular mode of the calculator to RADians. Then proceed as usual by entering the angle in terms of π. Angles that are expressed as fractional parts of π need to be enclosed in parentheses. For example, to find the value of $\sin \frac{\pi}{3}$, use one of the following key sequences:

Thus, correct to four decimal places, $\sin \frac{\pi}{3} \approx \mathbf{0.8660}$.

Reciprocal Functions

Secant, cosecant, and *cotangent* (abbreviated as *sec, csc,* and *cot*) are the names given to the functions that are the reciprocals of cosine, sine, and tangent, respectively. Thus, for all values of θ except those that make the denominator 0, the formulas for the reciprocal functions are as follows:

$$\sec \theta = \frac{1}{\cos \theta}, \quad \csc \theta = \frac{1}{\sin \theta}, \quad \text{and } \cot \theta = \frac{1}{\tan \theta}.$$

To find the value of the secant, cosecant, or cotangent of an angle, use a scientific calculator to find the value of the reciprocal function (cosine, sine, or tangent). Then find the reciprocal of the computed value in the calculator display window by using the calculator's reciprocal key, $\boxed{1/x}$. For example:

- Since $\sec 50° = \dfrac{1}{\cos 50°}$, you can find the value of $\sec 50°$ correct to four decimal places by using a key sequence similar to this one:

$$50 \;\boxed{\text{COS}}\; \boxed{1/x}.$$

Thus, $\sec 50° \approx \mathbf{1.5557}$.

- Since $\csc 35°20' = \dfrac{1}{\sin 35°20'}$, you can find the value of $\csc 35°20'$ correct to four decimal places by using a key sequence similar to this one:

$$35.3333 \boxed{\text{SIN}} \boxed{1/x}.$$

Thus, $\csc 35°20' \approx \mathbf{1.7291}$.

Coordinate Definitions of the Six Trigonometric Functions

The six trigonometric functions of an angle in standard position can be defined in terms of the coordinates of *any* point $P(x, y)$ on its terminal side and its distance r from the origin. In the accompanying chart, it is understood that x and y cannot be 0 when either appears in the denominator of a fraction.

Function		Reciprocal
$\sin \theta = \dfrac{y}{r}$	and	$\csc \theta = \dfrac{r}{y}$
$\cos \theta = \dfrac{x}{r}$	and	$\sec \theta = \dfrac{r}{x}$
$\tan \theta = \dfrac{y}{x}$	and	$\cot \theta = \dfrac{x}{y}$

Since these definitions do not depend on a right triangle, trigonometric functions of angles *greater* than 90° may be determined if the values of x, y, and r are known. Notice that, as the terminal side rotates 360°, $P(x, y)$ determines a circle whose center is at the origin and whose radius is $OP = r$. Thus the values of x, y, and r must satisfy the relationship

$$x^2 + y^2 = r^2.$$

Example

4. If $P(\sqrt{7}, -3)$ lies on a terminal ray of an angle θ, find the exact value of each of the six trigonometric functions of θ.

Solution: Since $P(\sqrt{7}, -3) = P(x, y)$, let $x = \sqrt{7}$ and $y = -3$.

Since $x > 0$ and $y < 0$, $\angle\theta$ terminates in Quadrant IV.

Find r:

$$x^2 + y^2 = r^2$$
$$(\sqrt{7})^2 + (-3)^2 = r^2$$
$$7 + 9 = r^2$$
$$16 = r^2$$

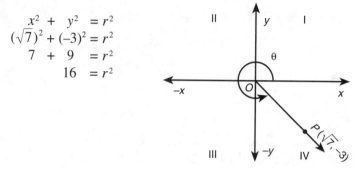

Since r is always positive, $r = +\sqrt{16} = 4$. Applying the coordinate definitions where $x = \sqrt{7}$, $y = -3$, and $r = 4$ gives:

- $\sin \theta = \dfrac{y}{r} = -\dfrac{3}{4}$ and $\csc \theta = \dfrac{r}{y} = \dfrac{4}{-3} = -\dfrac{4}{3}$

- $\cos \theta = \dfrac{x}{r} = \dfrac{\sqrt{7}}{4}$ and $\sec \theta = \dfrac{r}{x} = \dfrac{4}{\sqrt{7}}$

- $\tan \theta = \dfrac{x}{y} = -\dfrac{3}{7}$ and $\cot \theta = \dfrac{x}{y} = \dfrac{\sqrt{7}}{-3} = -\dfrac{\sqrt{7}}{3}$

Finding the Signs of Trigonometric Functions

If $P(x, y)$ is any point of the terminal side of an angle θ in standard position $(x, y \neq 0)$, then the values of the trigonometric functions of $\angle\theta$ are positive or negative, depending on the signs of x and y in the quadrant in which the angle lies. The possibilities are listed in the accompanying table. Notice that Quadrant I is the only quadrant in which all of the six trigonometric functions are positive at the same time. In each of the other quadrants, only one of the basic trigonometric functions and its reciprocal function are positive.

Trigonometric Function	Quadrant in Which θ Lies			
	I	II	III	IV
$\sin \theta$ and $\csc \theta$	+	+	−	−
$\cos \theta$ and $\sec \theta$	+	−	−	+
$\tan \theta$ and $\cot \theta$	+	−	+	−

The first letter in each word of the sentence

$$\textbf{A}ll \ \textbf{S}tudents \ \textbf{T}ake \ \textbf{C}alculus$$

is helpful in remembering the quadrants in which the different trigonometric functions are positive. Note that:

$A = A$ll functions are positive in Quadrant I.
$S = S$ine and cosecant are positive in Quadrant II.
$T = T$angent and cotangent are positive in Quadrant III.
$C = C$osine and secant are positive in Quadrant IV.

Example

5. If $\cos \theta = -\dfrac{4}{5}$ and $\tan \theta$ is positive, what is the value of $\sin \theta$?

Solution: Cosine is negative in the second and *third* quadrants. Tangent is positive in the first and *third* quadrants. Hence in Quadrant III cosine is negative and, at the same time, tangent is positive.

Let $P(x, y)$ be a point on the terminal side of $\angle \theta$ such that, if $\cos \theta = -\dfrac{4}{5} = \dfrac{x}{r}$,

then $x = -4$ and $r = 5$. Find y:

$$x^2 + y^2 = r^2$$
$$(-4)^2 + y^2 = 5^2$$
$$16 + y^2 = 25$$
$$y^2 = 9$$
$$y = \pm\sqrt{9} = \pm 3$$

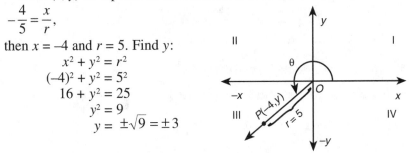

Since $\angle \theta$ terminates in Quadrant III, $y = -3$. Thus, $x = -4$, $y = -3$, $r = 5$, and

$$\sin \theta = \frac{y}{r} = -\frac{3}{5}.$$

Cofunction Relationships

The use of the prefix *co-* in *cosine, cosecant,* and *cotangent* is not accidental. The prefix represents "complementary." Each of these functions has the same value as its "complementary" function when the angles are complementary. Thus:

$\sin \theta = \cos (90 - \theta)$	*Examples:* *1.* $\sin 60°$ $= \cos 30°$
$\sec \theta = \csc (90 - \theta)$	*2.* $\sec 15°$ $= \csc 75°$
$\tan \theta = \cot (90 - \theta)$	*3.* $\tan 20°50′ = \cot 69°10′$

Exercise Set 7.2

1. Change from degrees to radians:
   (a) 90°   (c) 330°   (e) –225°
   (b) 100°   (d) 198°   (f) 120°

2. Change from radians to degrees:
   (a) $\dfrac{\pi}{3}$ (b) $\dfrac{4}{9}\pi$ (c) $\dfrac{7}{6}\pi$ (d) $-\dfrac{5}{12}\pi$ (e) $\dfrac{7}{4}\pi$ (f) -2π

3. Find the number of radians in a central angle of a circle that intercepts an arc of 2 feet if the radius of the circle is 3 feet.

4. In a circle whose radius is 4 inches, find the length of an arc intercepted by a central angle of 2 radians.

5. Find x correct to four decimal places:
   (a) cot 50°   (b) csc 72°30′   (c) sec 19°40′

6. Find x correct to the *nearest 10 minutes* if:

   (a) sec $x = 3.1708$ (c) csc $x = 1.0790$ (e) sec $x = \dfrac{\pi}{2}$

   (b) cot $x = 1.6971$ (d) cot $x = \pi$ (f) csc $x = \dfrac{\pi}{3}$

7. Find the number of inches in the radius of a circle in which a central angle of 2.5 radians intercepts an arc of 15 inches.

8. Find the number of inches in the length of an arc of a circle intercepted by a central angle of $\dfrac{1}{2}$ radian, if the circumference of the circle is 8π inches.

9. If point $P(-2, \sqrt{5})$ lies on the terminal ray of obtuse angle θ, find the value of each of the six trigonometric functions.

10. Find the exact value of each of the five other trigonometric functions if:
    (a) tan $\theta = -\dfrac{5}{12}$ and sin θ is negative

    (b) sin $\theta = -\dfrac{1}{\sqrt{5}}$ and cos θ is negative

11. Find the exact value of cos θ if cot $\theta = \dfrac{-15}{8}$ and csc θ is positive.

12. Find the exact value of sin θ if tan $\theta = 4$ and cos θ is negative.

13. A ball is rolling in a circular path that has a radius of 10 inches. What distance has the ball rolled when the subtended arc is 54°? Express your answer to the *nearest hundredth of an inch*.

14 and 15. Find the value of positive acute angle x.

14. $\tan (2x)° = \cot (3x)°$

15. $\sin (3x + 28)° = \cos (2x + 12)°$

16. Through how many radians does the minute hand of a clock turn in 24 minutes?

7.3 WRITING TRIGONOMETRIC FUNCTIONS USING REFERENCE ANGLES

KEY IDEAS

In working with trigonometric functions, it is useful to be able to express the functions of an angle in Quadrant II, III, or IV as trigonometric functions of a positive acute angle.

Finding Reference Angles

For any angle θ in standard position, the **reference angle** is the *acute* angle θ_R, whose vertex is the origin and whose sides are the terminal side of $\angle\theta$ and the *x*-axis. For a Quadrant I angle, the original angle and the reference angle are the same. The reference angles in the other quadrants are shown in Figure 7.5.

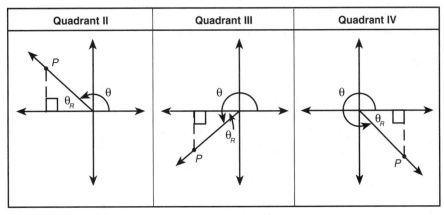

Quadrant II	Quadrant III	Quadrant IV

Figure 7.5 Reference Angles in Quadrants II, III, IV

Reducing Angles of Trigonometric Functions

The trigonometric function of *any* angle θ can be expressed as either plus or minus the same trigonometric function of its reference angle, θ_R. For example, to reduce cos 135°:

- Find the reference angle. As shown in Figure 7.6, $\theta_R = 45°$.
- Determine the sign of the consine function in the quadrant in which θ_R is located. In Quadrant II, cosine is negative. Hence

$$\cos 135° = -\cos 45°.$$

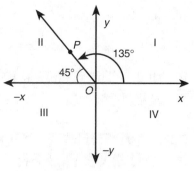

Figure 7.6   Reference Angles

Because sine is positive in Quadrant II while tangent is negative in Quadrant II:

$$\sin 135° = \sin 45° \quad \text{and} \quad \tan 135° = -\tan 45°.$$

Sometimes it is necessary to reduce functions of angles greater than 360° or less than 0°. You should draw a diagram and verify that:

- $\cos 570° = \cos(570 - 360)° = \cos 210° = -\cos 30°.$
- $\sin(-140)° = \sin 220° = -\sin 40°.$
- $\tan 650° = \tan 290° = -\tan 70°.$

Example

1. Express as an equivalent function of a positive acute angle:

   (a) tan 130°   **(b)** sin (–110°)   **(c)** sec 485°   **(d)** cos $\dfrac{7\pi}{9}$

Solutions: **(a)** An angle of 130° terminates in Quadrant II. The reference angle is 180° – 130° or 50°. Since tangent is negative in Quadrant II,

$$\tan 130° = -\tan 50°.$$

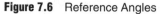

(b) An angle of –110° represents a *clockwise* rotation of 110°, so that the terminal side of the angle lies in Quadrant III, forming an angle of 70° with the negative x-axis. Since $\theta_R = 70°$ and sine is negative in Quadrant III,

$$\sin(-110°) = -\textbf{sin 70°}.$$

(c) An angle of 485° is coterminal with an angle of 485° – 360° or 125°, which lies in Quadrant II. Since the reference angle is 180° – 125° or 55°, and secant is negative in Quadrant II,

sec 485° = sec 125° = **–sec 55°**.

(d) Since $\dfrac{7\pi}{9} > \dfrac{\pi}{2}$ but $\dfrac{7\pi}{9} > \pi$,

θ lies in Quadrant II. Hence

$$\theta_R = \pi - \frac{7\pi}{9} = \frac{9\pi}{9} - \frac{7\pi}{9} = \frac{2\pi}{9}.$$

Since cosine is negative in the second quadrant,

$$\cos\frac{7\pi}{9} = -\textbf{cos}\frac{2\pi}{9}.$$

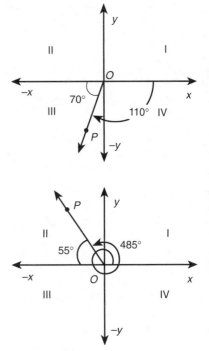

Functions of Special Angles

Finding the exact values of trigonometric functions of angles measuring 30°, 60°, and 45° is sometimes required. The values of the trigonometric functions of 30° and 60° are summarized in the following table, which should be memorized:

x	Sin x	Cos x	Tan x
30°	$\dfrac{1}{2}$	$\dfrac{\sqrt{3}}{2}$	$\dfrac{1}{\sqrt{3}}$ or $\dfrac{\sqrt{3}}{3}$
60°	$\dfrac{\sqrt{3}}{2}$	$\dfrac{1}{2}$	$\sqrt{3}$

The next table summarizes values of trigonometric functions of 45°, which should also be memorized:

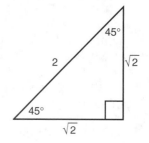

x	Sin x	Cos x	Tan x
45°	$\dfrac{\sqrt{2}}{2}$	$\dfrac{\sqrt{2}}{2}$	1

Example

2. Find the exact value:

(a) cos 120°   **(b)** tan 225°   **(c)** sin 300°   **(d)** cos (−30°)

Solutions: In each case, you should first determine the quadrant in which the angle terminates. If a diagram will help, draw it.

(a) An angle measuring 120° terminates in Quadrant II. The measure of the reference angle is 180° − 120° = 60°. Since cosine is negative in the second quadrant,

$$\cos 120° = -\cos 60° = -\frac{1}{2}.$$

(b) An angle measuring 225° terminates in Quadrant III. Since θ_R = 225° − 180° = 45° and tangent is positive in the third quadrant,

$$\tan 225° = \tan 45° = \mathbf{1}.$$

(c) An angle measuring 300° terminates in Quadrant IV. Since θ_R = 360° − 300° = 60° and sine is negative in the fourth quadrant,

$$\sin 300° = -\sin 60° = -\frac{\sqrt{3}}{2}.$$

(d) An angle of −30° represents a *clockwise* rotation of 30°, so the terminal side of the angle lies in Quadrant IV, forming an angle of 30° with the positive *x*-axis. Since θ_R = 30° and cosine is positive in Quadrant IV,

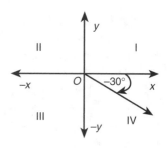

$$\cos (-30°) = \cos 30° = \frac{\sqrt{3}}{2}.$$

Quadrantal Angles

An angle whose terminal side coincides with a coordinate axis, as in Figure 7.7, is called a **quadrantal angle.** An angle of rotation whose measure is 0°,

90°, 180°, 270°, or 360° is a *quadrantal* angle. If $P(x, y)$ is on the terminal side of an angle of rotation of 90°, then $P(x, y)$ falls on the positive y-axis so that $x = 0$ and $y = r$. Hence

$$\sin 90° = \frac{y}{r} = \frac{r}{r} = 1$$

and

$$\cos 90° = \frac{x}{r} = \frac{0}{r} = 0.$$

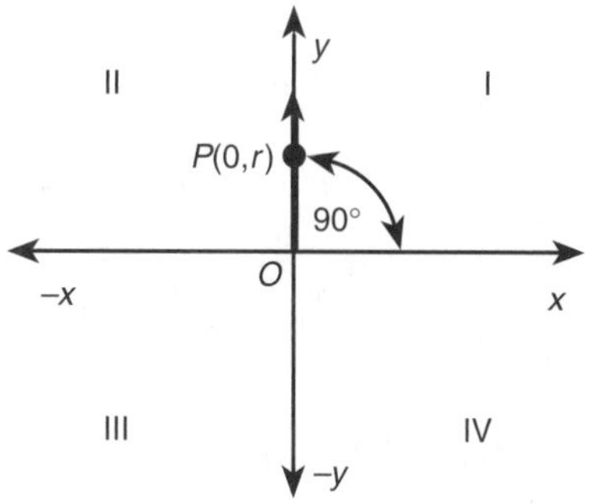

Figure 7.7 Quadrantal Angles

Since $\tan 90° = y \div x$ and $x = 0$, $\tan 90°$ is not defined.

A similar analysis makes it possible to find the values of the other quadrantal angles listed in Table 7.1. By inverting the values of sine, cosine, and tangent, the values of the corresponding reciprocal functions (cosecant, secant, and cotangent) can be determined, provided that a zero denominator does not result. For example, since

$$\sec 90° = \frac{1}{\cos 90°} = \frac{1}{0},$$

sec 90° is not defined.

You can also use your scientific calculator to find the values of the sine, cosine, and tangent of the quadrantal angles, provided that they exist.

TABLE 7.1 EVALUATING FUNCTIONS OF QUADRANTAL ANGLES

Trig Function	$\frac{\pi}{2}$	π	$\frac{3\pi}{2}$	2π
	$\theta = 90°$	$\theta = 180°$	$\theta = 270°$	$\theta = 360°$ (or 0°)
Sin θ	1	0	−1	0
Cos θ	0	−1	0	1
Tan θ	Undefined	0	Undefined	0

Note: Since 0° and 360° (2π) are coterminal angles, they are listed together.

Example

3. If $f(x) = 2 \sin x + \cos 2x$, find $f\left(\frac{\pi}{2}\right)$.

Solution: Let $x = \frac{\pi}{2}$. Then:

$$f\left(\frac{\pi}{2}\right) = 2 \sin \frac{\pi}{2} + \cos 2\left(\frac{\pi}{2}\right)$$

$$= 2(1) \qquad + \cos \pi$$
$$= 2 \qquad + (-1)$$
$$= 1$$

Exercise Set 7.3

1–8. Express each as an equivalent function of a positive acute angle.

1. $\cos 105°$ **3.** $\sin 250°$ **5.** $\sin \dfrac{5\pi}{3}$ **7.** $\tan \dfrac{5\pi}{4}$

2. $\tan (-110°)$ **4.** $\csc 490°$ **6.** $\cos \dfrac{7\pi}{6}$ **8.** $\sec 505°$

9–20. Find the exact value.

9. $\sin 120°$ **12.** $\cos 300°$ **15.** $\cos 135°$ **18.** $\sin 300°$

10. $\tan 315°$ **13.** $\csc (-30°)$ **16.** $\tan 150°$ **19.** $\cot (-135°)$

11. $\sin \dfrac{4}{3}\pi$ **14.** $\cot \dfrac{5\pi}{4}$ **17.** $\sec \dfrac{7\pi}{6}$ **20.** $\csc \dfrac{11\pi}{3}$

21–26. If $0° \le x < 360°$, find possible values for x that satisfy each equation.

21. $\sin x = \dfrac{\sqrt{2}}{2}$ **23.** $\tan x = -1$ **25.** $\cot x = \sqrt{3}$

22. $\sin x = -\dfrac{1}{2}$ **24.** $\cos x = -\dfrac{\sqrt{3}}{2}$ **26.** $\sec x = 2$

27 and 28. In each of the following, find $f\left(\dfrac{\pi}{4}\right)$.

27. $f(x) = 2 \tan x - \sin 2x$

28. $f(x) = \cot x - \tan 3x$

29 and 30. In each of the following find $f\left(\dfrac{\pi}{2}\right)$.

29. $f(x) = 2 \sin 3x + \cos x$

30. $f(x) = \cos 4x - \cos 2x + \sin x$

31–35. If $0° \le x < 360°$, for which value of x is the given expression not defined?

31. $\sec x$ **32.** $\csc x$ **33.** $\dfrac{1}{1+\sin x}$ **34.** $\dfrac{1}{1+\tan x}$ **35.** $\dfrac{1}{1-2\cos x}$

7.4   SKETCHING $y = a \sin bx$ and $y = a \cos bx$

⚠ **KEY IDEAS**

The function values of sine and cosine repeat every 360°. For example,

$$\sin 40° = \sin (40 + 360)° = \sin (40 + 720)° = \ldots$$

As a result, sine and cosine are examples of *periodic functions.* The graphs of the sine and cosine functions have fundamentally the same wavelike shape, which repeats every 360° or 2π radians. Each repetition of the graph is called a *cycle.*

Comparing Sine and Cosine Curves

Table 7.2 and 7.3 summarize the key features of the graphs of $y = \sin x$ and $y = \cos x$. The sine and cosine curves have the same basic shape, but the cosine curve is translated horizontally 90°.

The sine and cosine functions have the same *amplitude, period,* and *frequency.* The **amplitude** of $\sin x$ and $\cos x$ is their maximum value, which is 1. The **period** of the sine and cosine curves is 2π radians or 360°, since their shapes are repeated every 2π radians. In an interval of 2π radians, the graphs of $\sin x$ and $\cos x$ each complete one full cycle, so their **frequency** is 1.

TABLE 7.2 FEATURES OF $y = \sin x\ (-2\pi \le x \le 2\pi)$

Feature	$y = \sin x$	Graph $(-2\pi \le x \le 2\pi)$
Amplitude	1	
Frequency	1	
Period	2π	
y-Intercept	$y = 0$	
x-Intercepts	multiples of π	
Symmetric to:	origin	

177

TABLE 7.3 FEATURES OF $y = \cos x$ $(-2\pi \le x \le 2\pi)$

Feature	$y = \cos x$	Graph $(-2\pi \le x \le 2\pi)$
Amplitude	1	
Frequency	1	
Period	2π	
y-Intercept	$y = 1$	
x-Intercepts	odd multiples of $\dfrac{\pi}{2}$	
Symmetric to:	y-axis	

Examples

1. On the interval $0 \le x \le 2\pi$, sketch the graphs of $y = \sin x$ and $y = \cos x$ on the same set of axes. In which quadrant is $\sin x$ decreasing and $\cos x$ increasing?

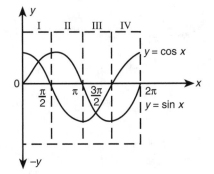

Solution: In Quadrant III, the graph of $y = \sin x$ falls while the graph of $y = \cos x$ rises. Thus $\sin x$ is decreasing and $\cos x$ is increasing in Quadrant **III**.

2. On the interval $-\pi \le x \le \pi$, sketch the graph of $y = \sin x$ and its image under the transformation $T_{\pi,0}$.

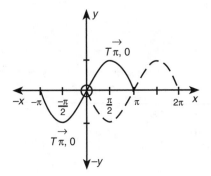

Solution: The solid curve is the graph of $y = \sin x$ on the interval $-\pi \le x \le \pi$. The broken curve is the image of $y = \sin x$ under the transformation $T_{\pi,0}$, which translates the graph in the horizontal direction π radians to the right.

Curves with Different Amplitudes and Periods

In an equation of the form

$$y = a \sin bx \quad \text{or} \quad y = a \cos bx,$$

178

a multiplies the value of the function so that *a* determines the maximum height of amplitude of the graph. The value of *b* multiplies the angle *x* so that its value affects the frequency with which the graph repeats itself in an interval of 2π radians. If $y = a \sin bx$ or $y = a \cos bx$, then

$$\text{amplitude} = |a|, \quad \text{frequency} = |b|, \quad \text{period} = \frac{2\pi}{|b|}.$$

To sketch the graph of $y = 3 \sin 2x$ on $0 \le x \le 2\pi$:

- *Determine the amplitude and the period.* For the equation $y = 3 \sin 2x$, $a = 3$ and $b = 2$. Therefore the amplitude is 3, and the period is $\frac{2\pi}{2}$ or π radians.
- *Frame the graph, using rectangles.* Since the period is π radians, the graph will complete one cycle from 0 to π radians and another full cycle from π to 2π radians. Label the *x*-axis with these points. You may also reason that, since the coefficient of *x* in $y = 3 \sin 2x$ is 2, the frequency is 2, so that the curve completes *two* full cycles in an interval of 2π radians.

  The amplitude is 3, so the graph does not rise above $y = 3$ or fall below $y = -3$. Indicate this fact by drawing two horizontal boundary lines through $y = 3$ and $y = -3$, stopping at $x = 2\pi$. Then form two rectangles by drawing vertical lines through $x = \pi$ and $x = 2\pi$. Each of these rectangles will contain one complete cycle of the graph. Next, divide the *x*-axis from 0 to π into four intervals of equal length, and do the same for the interval from π to 2π. At these key points, the curve either passes through the *x*-axis, achieves a maximum height, or reaches a minimum height.
- *Sketch the curve.* Knowing the basic shape of the sine curve, draw one complete cycle from 0 to π radians, starting at $(0, 0)$ (see Figure 7.8). Use the points marked off on the *x*-axis and the rectangular boundary lines as guideposts. Then draw another complete cycle from π to 2π radians.

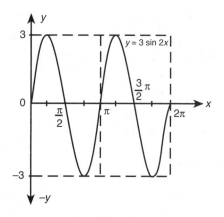

Figure 7.8 Graph of $y = 3 \sin 2x$ ($0 \le x \le 2\pi$)

Examples

3. Determine the equation of each graph:

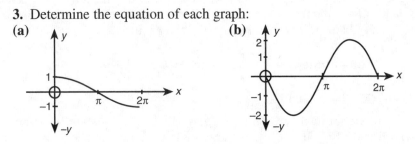

(a)

(b)

Solutions: **(a)** The curve has the basic shape of a cosine curve whose amplitude is 1. The curve completes one-half of a complete cycle in 2π radians, so its frequency is $\frac{1}{2}$. Since $a = 1$ and $b = \frac{1}{2}$, the equation of the graph is

$$y = \cos \frac{1}{2}x.$$

(b) The curve has the basic shape of a sine curve whose amplitude is 2, except that it has been reflected in the *x*-axis. The equation of the original curve is, therefore, $y = 2 \sin x$. To reflect the graph of this equation in the *x*-axis, replace *y* by $-y$, resulting in $-y = 2 \sin x$ or

$$y = -2 \sin x.$$

4. (a) On the same set of axes, sketch and label the graph of the equations $y = 2 \sin \frac{1}{2}x$ and $y = \cos 2x$ in the interval $0 \le x \le 2\pi$.

(b) What is a value of *x* in the interval $0 \le x \le 2\pi$ for which $2 \sin \frac{1}{2}x - \cos 2x = 1$?

Solutions: **(a)** The equation $y = 2 \sin \frac{1}{2}x$ has the form $y = a \sin bx$, where $a = 2$ and $b = \frac{1}{2}$. Hence the amplitude of this graph is 2, and its frequency is $\frac{1}{2}$. Since the amplitude is 2, the sine curve achieves a maximum height of 2 and reaches a minimum height of -2. A frequency of $\frac{1}{2}$ means that in the interval specified by the question, 0 to 2π, the curve will complete one-half of a full cycle. This represents one arch of the sine curve, which appears in the accompanying graph.

The equation $y = \cos 2x$ has the form $y = a \cos bx$, where $a = 1$ and $b = 2$. The amplitude of the curve is 1, and its frequency is 2.

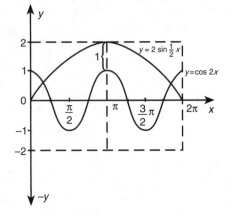

The curve reaches a maximum height of 1 and a minimum height of –1. Mark off on the x-axis four intervals of equal length from 0 to π radians, and four intervals of equal length from π to 2π radians. Knowing the basic shape of the cosine curve, draw one complete cycle from 0 to π radians, and another complete cycle from π to 2π radians.

(b) For $2 \sin \frac{1}{2}x - \cos 2x = 1$, the height of the sine curve must be exactly one unit greater than the height of the cosine curve. Looking at the curves, you see that at $x = \pi$ the height of $y = 2 \sin \frac{1}{2}x$ is 2 and the height of $y = \cos 2x$ is 1. Therefore

$$2 \sin \frac{1}{2}x - \cos 2x = 1 \quad \text{at } x = \pi.$$

Exercise Set 7.4

1. Write an equation of the cosine curve if the amplitude is 3 and the frequency is:

    (a) 1   **(b)** 2   **(c)** $\frac{1}{2}$

2. Write an equation of the sine curve if the amplitude is 2 and the period is:

    (a) 2π **(b)** π **(c)** 4π

3. Write an equation of the horizontal line that is above the x-axis and tangent to the graph of:

    (a) $y = 3 \sin 2x$ **(b)** $y = 2 \cos \frac{1}{2}x$

4. If $0 \le x < 2\pi$, for what value of x does the graph of each function reach its *maximum* (highest point)?

    (a) $y = 3 \sin \frac{1}{2}x$ **(b)** $y = -\cos 2x$

5. If $y = 2 \cos 3x$, what is the *minimum* value of y?

6. Determine the number of times the graph of $y = \sin 2x$ crosses the x-axis as x increases from 45° to 315°.

7. In which quadrant is ∠x located if the graphs of $y = \sin x$ and $y = \cos x$ are both decreasing when ∠x is increasing?

8. The graph of which function passes through the point whose coordinates are $(\pi, 2)$?

(1) $y = 2 \cos \dfrac{x}{2}$ (3) $y = 2 \sin \dfrac{x}{2}$

(2) $y = \dfrac{1}{2} \cos 2x$ (4) $y = \dfrac{1}{2} \sin 2x$

9. The graph of which equation has a period of π radians and passes through the origin?

(1) $y = \cos 2x$ (2) $y = \sin 2x$ (3) $y = \cos \dfrac{x}{2}$ (4) $y = \sin \dfrac{x}{2}$

10. As θ increases from π to 2π radians, the graph of $y = \sin \theta$
    (1) increases throughout the interval
    (2) decreases throughout the interval
    (3) decreases, then increases
    (4) increases, then decreases

11. **(a)** On the same set of axes, draw and label the graphs of $y = \sin \dfrac{1}{2}x$ and $y = 2 \cos x$, as x varies from 0 to 2π radians.
    (b) From the graphs made in answer to part **(a)**, find the number of values of x greater than 0 and less than 2π for which $2 \cos x - \sin \dfrac{1}{2}x = 0$.

12. **(a)** On the same set of axes, draw and label the graphs of $y = 2 \sin x$ and $y = \cos 2x$ in the interval $-\pi \le x \le \pi$.
    (b) Answer the following from the graph made in part **(a)**:
    (1) What is the value of the difference $2 \sin x - \cos 2x$ at $x = \dfrac{x}{2}$?
    (2) How many values in the interval $-\pi \le x \le \pi$ satisfy the equation $2 \sin x - \cos 2x = 0$?

13. **(a)** On the same set of axes, draw and label the graphs of the equations $y = 2 \sin x$ and $y = \cos \dfrac{1}{2}x$ for the values in the interval $0 \le x \le 2\pi$.

    (b) Each graph drawn in part **(a)** is symmetric about which of the following?
    (1) line $x = \pi$ (2) x-axis   (3) point $(\pi, 0)$ (4) origin

14–16. On the same set of axes, sketch and label the graph of each function and its image under the given transformation.

	Function	Interval	Transformation
14.	$y = \cos x$	$0 \le x \le 2\pi$	$r_{x\text{-axis}}$
15.	$y = 2 \sin x$	$0 \le x \le 2\pi$	$r_{x\text{-axis}}$
16.	$y = 2 \cos \dfrac{1}{2} x$	$-2\pi \le x \le 2\pi$	$D_{\frac{1}{2}}$

7.5 SKETCHING $y = $ TAN x

===== KEY IDEAS =====

The graph of $y = \tan x$, unlike the sine and cosine curves, has no amplitude, has breaks in continuity, and repeats every π radians rather than every 2π radians.

Basic Features of the Tangent Curve

The function $y = \tan x$ is not defined at $x = 90°$ $\left(=\frac{\pi}{2}\right)$ or at $x = 270°$ $\left(=\frac{3\pi}{2}\right)$ or at any other odd multiple of $\pm \frac{\pi}{2}$. As x approaches an odd multiple of $\pm \frac{\pi}{2}$, the value of $\tan x$ becomes unbounded. The graph of the tangent function, shown in Figure 7.9, has these special features:

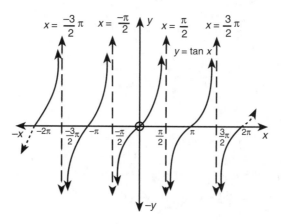

Figure 7.9   Graph of $y = \tan x$ ($-2\pi \le x \le 2\pi$)

183

- The graph has no amplitude. Instead, vertical lines through odd multiples of $\pm \frac{\pi}{2}$ serve as asymptotes to the graph.
- The *period* of the graph is π radians, with the graph completing one full cycle between consecutive pairs of vertical asymptotes.
- The graph has x-intercepts at integer multiples of π.

Exercise Set 7.5

1. As x increases from $0°$ to $360°$, the graph of $y = \tan x$ rises in
   (1) no quadrant
   (2) all four quadrants
   (3) the first and third quadrants, only
   (4) the second and fourth quadrants, only

2. (a) On the same set of axes, sketch the graphs of $y = 3 \sin x$ and $y = \tan x$ as x varies from $-\pi$ to π radians.
   (b) From the graphs made in answer to part (a), determine the number of roots of $3 \sin x - \tan x = 0$ that lie between $-\frac{\pi}{2}$ and $\frac{\pi}{2}$ radians.

3. (a) On the same set of axes, sketch the graphs of $y = \cos 2x$ and $y = \tan x$ as x varies from $-\frac{\pi}{2}$ to $\frac{\pi}{2}$ radians.
   (b) From the graphs made in answer to part (a), determine the number of points between $-\frac{\pi}{2}$ and $\frac{\pi}{2}$ radians for which $\tan x - \cos 2x = 0$.

7.6 FINDING INVERSE TRIGONOMETRIC FUNCTIONS

∧ KEY IDEAS ∠

Since the graphs of

$$y = \sin x, \quad y = \cos x, \quad \text{and} \quad y = \tan x$$

do not pass the horizontal-line test, their inverses are not functions. If, however, the domains of the original trigonometric functions are restricted so that their graphs pass the horizontal-line test, then their inverses are functions.

Using Inverse Trigonometric Notation

To form the inverse of $y = \sin x$, interchange x and y to obtain $x = \sin y$. Since there is no algebraic method that can be used to solve for y in terms of x, a new notation must be introduced.

If $y = \sin x$, then its inverse is written as

$$y = \textbf{arc sin } x.$$

When you read "arc sin x" think "the *angle* whose sine has the value x." For example, if $y = \text{arc sin } \frac{1}{2}$, then y is the angle whose sine is one-half. Since $\sin 30° = \sin 150° = \frac{1}{2}$, then $y = 30°$ or $150°$; that is, the ordered pairs $(\frac{1}{2}, 30°)$ and $(\frac{1}{2}, 150°)$ satisfy the inverse relation. The inverse is not a function since the same value of x is paired with two different values of y.

The expression arc sin x may also be written, using inverse function notation, as $\sin^{-1} x$. Do not confuse the inverse notation $\sin^{-1} x$ with $(\sin x)^{-1}$, which means $\frac{1}{\sin x}$, or with $\sin x^{-1}$, which means $\sin \frac{1}{x}$.

Similarly, the inverse of $y = \cos x$ is

$$y = \textbf{arc cos } x \quad \text{or} \quad y = \cos^{-1} x,$$

and the inverse of $y = \tan x$ is

$$y = \textbf{arc tan } x \quad \text{or} \quad y = \tan^{-1} x.$$

Example

1. In the interval $0 \le \theta < 2\pi$, find:

(a) arc cos $\dfrac{1}{2}$ **(b)** arc tan (-1)

Solutions: **(a)** Finding arc cos $\dfrac{1}{2}$ means finding the angle (or angles) whose cosine is $\dfrac{1}{2}$. In the interval $0 \le \theta < 2\pi$, cosine is positive in Quadrants I and IV, so that $\cos 60° = \cos 300° = \dfrac{1}{2}$. Hence

$$\text{arc cos } \frac{1}{2} = \textbf{60}° \text{ or } \textbf{300}°.$$

(b) Finding arc tan (-1) means finding the angle (or angles) whose tangent is -1. In the interval $0 \le \theta < 2\pi$, tangent is negative in Quadrants II and IV, so that $\tan 135° = \tan 315° = -1$. Hence

$$\text{arc tan } (-1) = \textbf{45}° \text{ or } \textbf{315}°.$$

Forming Inverse Trigonometric Functions

Figure 7.10 shows that in the interval $-\frac{\pi}{2} \le x \le \frac{\pi}{2}$ the graph of $y = \sin x$ passes the horizontal-line test while $\sin x$ takes on its full range of values from -1 to 1. Hence, if the domain of $y = \sin x$ is restricted to $-\frac{\pi}{2} \le x \le \frac{\pi}{2}$, then its inverse is a function on this interval and has the restricted range $-\frac{\pi}{2} \le y \le \frac{\pi}{2}$.

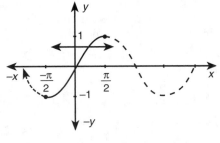

Figure 7.10 Graph of $y = \sin x \left(-\frac{\pi}{2} \le x \le \frac{\pi}{2}\right)$

The angle measures in the restricted range are called **principal values.** The principal values of the inverse *function* for sine are indicated by capitalizing the first letter of arc sin x or $\sin^{-1} x$, as in Arc sin x or Sin^{-1} x.

As is shown in Table 7.4, the domains of $y = \cos x$ and $y = \tan x$ may also be restricted so that their inverses are functions. Notice that the principal values (range) of Arc sin x and Arc tan x are restricted to the first and *fourth* quadrants, while the principal values (range) of Arc cos x are in the first and *second* quadrants.

TABLE 7.4 INVERSE TRIGONOMETRIC FUNCTIONS

Original Function	Restricted Domain	Inverse Function	Principal Values
$y = \sin x$	$-90° \le x \le 90°$	$y = $ Arc sin x	$-90° \le y \le 90°$
$y = \cos x$	$0° \le x \le 180°$	$y = $ Arc cos x	$0° \le y \le 180°$
$y = \tan x$	$-90° < x < 90°$	$y = $ Arc tan x	$-90° < y < 90°$

Examples

2. Find:

(a) Arc cos $\dfrac{1}{2}$ **(b)** Arc cos $\left(-\dfrac{1}{2}\right)$

Solution: **(a)** Let $\theta = $ Arc cos $\dfrac{1}{2}$. Then $\cos \theta = \dfrac{1}{2}$. The reference angle is 60°. Since the principal value of θ is restricted to Quadrants I and II and cosine is positive in Quadrant I, $\theta = 60°$. Hence

$$\text{Arc cos } \frac{1}{2} = 60°.$$

(b) Let $\theta =$ Arc cos $\left(-\dfrac{1}{2}\right)$. Then cos $\theta = -\dfrac{1}{2}$. Since the principal value of θ is restricted to Quadrants I and II, and cosine is negative in Quadrant II, $\theta = 120°$. Hence

$$\text{Arc cos}\left(-\frac{1}{2}\right) = \mathbf{120°}.$$

3. Find Arc tan $(-\sqrt{3})$.

Solution: Let $\theta =$ Arc tan $(-\sqrt{3})$. Then tan $\theta = \sqrt{3}$. The reference angle is $60°$. Since the principal value of θ is restricted to Quadrants I and IV, and tangent is negative in Quadrant IV, $\theta = -60°$. Hence

$$\text{Arc tan}\,(-\sqrt{3}) = \mathbf{-60°}.$$

4. Find:
(a) Arc csc 2   **(b)** Arc csc (–2)

Solutions: **(a)** Let $\theta =$ Arc csc 2. Then csc $\theta = 2$ or, equivalently, sin $\theta = \dfrac{1}{2}$. The reference angle is $30°$. Since the principal value of θ is restricted to Quadrants I and IV, $\theta = 30°$. Hence

$$\text{Arc csc } 2 = \mathbf{30°}.$$

(b) Let $\theta =$ Arc csc (–2). Then csc $\theta = $ -2, so that sin $\theta = -\dfrac{1}{2}$. Since the principal value of θ is restricted to Quadrants I and IV, and sine is negative in Quadrant IV, $\theta = -30°$. Hence

$$\text{Arc csc }(-2) = \mathbf{-30°}.$$

5. Find y:

(a) $y = \sin\,(\text{Arc tan } 1)$ **(b)** $y = \tan\left[\text{Arc cos}\left(-\dfrac{\sqrt{3}}{2}\right)\right]$

Solutions: Begin by evaluating the inverse function.
(a) Let $\theta =$ Arc tan 1. Then tan $\theta = 1$. The reference angle is $45°$. Since the principal value of θ lies only in Quadrants I and IV, $\theta = 45°$. Hence

$$y = \sin\,(\text{Arc tan } 1) = \sin 45° = \frac{\sqrt{2}}{2}.$$

(b) Let $\theta =$ Arc cos $\left(-\dfrac{\sqrt{3}}{2}\right)$. Then cos $\theta = -\dfrac{\sqrt{3}}{2}$. The reference angle is $30°$. The principal value of θ lies only in Quadrants I and II. Since cos θ is negative in Quadrant II, $\theta = 180° - 30° = 150°$. Thus

$$y = \tan\left[\text{Arc cos}\left(-\frac{\sqrt{3}}{2}\right)\right]$$

$$= \tan(150°) = -\tan 30°$$

$$= -\frac{\sqrt{3}}{3}$$

6. Find the value of $\cot\left[\text{Arc sin}\left(-\frac{3}{\sqrt{13}}\right)\right]$.

Solution: Let $\theta = \text{Arc sin}\left(-\frac{3}{\sqrt{13}}\right)$. Then $\sin\theta = -\frac{3}{\sqrt{13}} = \frac{y}{r}$. The principal value of θ lies only in Quadrants I and IV. Since $\sin\theta$ is negative in Quadrant IV, θ lies in Quadrant IV. The value of x can be determined from the relationship $x^2 + y^2 = r^2$ by letting $y = -3$ and $r = \sqrt{13}$:

$$x^2 + (-3)^2 = (\sqrt{13})^2$$
$$x^2 + 9 = 13$$
$$x^2 = 4$$
$$x = \pm 2$$

Since θ is in Quadrant IV, $x = 2$ and $\cot\theta = \frac{x}{y} = \frac{2}{-3}$ or $-\frac{2}{3}$. Thus

$$\cot\left[\text{Arc sin}\left(-\frac{3}{\sqrt{3}}\right)\right] = -\frac{2}{3}.$$

7. Find the value of $\text{Cos}^{-1}(-0.8515)$ correct to the:
(a) *nearest tenth of a degree*   **(b)** *nearest 10 minutes*

Solution: Here is a typical calculator key sequence for finding the number of degrees in the Quadrant II angle whose cosine is –0.8515:

0.8515   ±   2nd   COS

(a) Since $\text{Cos}^{-1}(-0.8515) \approx 148.375$, the unknown angle, correct to the *nearest tenth of a degree,* is 148.4°.
(b) Since there are 60 minutes per degree and

$$0.376 \text{ degree} \times 60 \frac{\text{minutes}}{\text{degree}} = 22.5 \text{ minutes,}$$

the unknown angle, correct to the *nearest 10 minutes,* is 148° 20′.

Exercise Set 7.6

1–10. Find the indicated angle correct to the: (a) nearest tenth of a degree *(b)* nearest 10 minutes.

1. Arc sin 1

2. Arc tan $\sqrt{3}$

3. Arc cos 0

4. Arc sin $\left(-\dfrac{1}{2}\right)$

5. $\cos^{-1}\dfrac{\sqrt{2}}{2}$

6. $\tan^{-1}$ (0.4699)

7. Arc sin (–1) – Arc cos (–1)

8. Arc tan $\left(-\dfrac{\sqrt{3}}{3}\right)$

9. $\sin^{-1}$ (–0.9520)

10. $\sec^{-1}$ (–1.4059)

11–19. Find the exact value.

11. Arc sin (–1)

12. Arc cos $\dfrac{\sqrt{3}}{2}$

13. Arc tan $(-\sqrt{3})$

14. Arc sec (–2)

15. Arc csc $\sqrt{2}$

16. Arc cot $\dfrac{\sqrt{3}}{3}$

17. $\sec^{-1} 1 + \csc^{-1} (-1)$

18. Arc sin $\dfrac{\sqrt{2}}{2}$ + Arc cos $\dfrac{\sqrt{2}}{2}$

19. $\tan^{-1} \sqrt{3} - \cot^{-1} (-\sqrt{3})$

20–36. Find the exact value.

20. sin (Arc cos 1)

21. cos (Arc tan 1)

22. $\tan\left(\text{Arc sin}\dfrac{\sqrt{2}}{2}\right)$

23. $\sin\left[\text{Arc cos}\left(-\dfrac{\sqrt{3}}{2}\right)\right]$

24. sin (Arc tan 3)

25. $\cos\left[\text{Sin}^{-1}\left(\dfrac{5}{13}\right)\right]$

26. $\cot\left(\text{Arc sin}\dfrac{3}{5}\right)$

27. $\tan\left(\text{Arc cot}\dfrac{1}{2}\right)$

28. tan (Arc tan $\sqrt{7}$)

29. $\tan\left[\text{Arc cos}\left(-\dfrac{24}{25}\right)\right]$

30. $\cos\left(\text{Arc sin}\dfrac{2}{3}\right)$

31. $\sin\left(\text{Arc tan}\dfrac{1}{3}\right)$

32. $\sec\left(\text{Arc cos}\dfrac{12}{13}\right)$

33. sec [Arc csc (–2)]

34. $\cos^{-1}\left(\sin\dfrac{7\pi}{6}\right)$

35. $\sin\left(\text{Arc sin}\dfrac{\sqrt{2}}{2}+\text{Arc cos}\dfrac{\sqrt{2}}{2}\right)$

36. cos [Arc cos (–1) + Arc tan $(-\sqrt{3})$]

189

REGENTS TUNE-UP: CHAPTER 7

Each of the questions in this section has appeared on a previous Course III Regents Examination. Here is an opportunity for you to review Chapter 7 and, at the same time, prepare for the Course III Regents Examination.

---

1. Express 3π radians in degrees.

2. If $\sin\theta = -\dfrac{4}{5}$ and θ is in Quadrant IV, find $\tan\theta$.

3. Express $\sin(-230°)$ as a function of a positive acute angle.

4. What is the exact value of $(\sin 60°)(\cos 60°)$?

5. As θ increases from π to $\dfrac{3\pi}{2}$, the value of $\cos\theta$

   (1) decreases, only   (3) decreases and then increases
   (2) increases, only   (4) increases and then decreases

6. What is the amplitude of the graph whose equation is $y = -4\sin 2x$?

7. The exact value of $\tan\left(\text{Arc}\cos\dfrac{\sqrt{2}}{2}\right)$ is

   (1) 1   (2) $\sqrt{3}$ (3) -1 (4) $\dfrac{\pi}{4}$

8. For which value of x is $\tan(x+20)°$ undefined?
   (1) -20 (2) 70   (3) 160   (4) 340

9. If $\cos(2x+10°) = \sin(x+20°)$, a value of x is
   (1) 20   (2) 30   (3) 40   (4) 60

10. If $f(x) = \cos\dfrac{x}{3} + \sin x$, then $f(\pi)$ equals

   (1) $\dfrac{3}{2}$ (2) $\dfrac{1}{2}$ (3) $-\dfrac{1}{2}$ (4) $\dfrac{\sqrt{3}}{2}$

11. What is the image of $(1, 0)$ after a counterclockwise rotation of $60°$?

   (1) $\left(\dfrac{1}{2}, \dfrac{\sqrt{3}}{2}\right)$ (2) $\left(\dfrac{\sqrt{3}}{2}, \dfrac{1}{2}\right)$ (3) $\left(\dfrac{1}{2}, -\dfrac{\sqrt{3}}{2}\right)$ (4) $\left(-\dfrac{\sqrt{3}}{2}, -\dfrac{1}{2}\right)$

12. If sec $x < 0$ and tan $x < 0$, then the terminal side of $\angle x$ is located in Quadrant
(1) I   (2) II   (3) III   (4) IV

13. Which field property is illustrated by the expression $(\tan \theta)(\cot \theta) = 1$?
(1) closure   (2) identity   (3) commutative   (4) inverse

14. The maximum value of $3 \sin \dfrac{1}{3}\theta$ is

(1) 1   (2) $\dfrac{1}{3}$ (3) 3   (4) 0

15. The graph of the equation $y = 2 \cos 2x$, $0 \le x \le 2\pi$, has a line of symmetry at

(1) $x = \pi$ (2) $x = \dfrac{\pi}{4}$ (3) $y = 2$ (4) the x-axis

16. **(a)** On the same set of axes, sketch and label the graphs of $y = \sin \dfrac{1}{2}x$ and $y = 2 \cos x$ as x varies from 0 to 2π radians.

(b) Using the same set of axes, sketch the reflection of $y = \sin \dfrac{1}{2}x$ in the line $y = -1$.

17. **(a)** On the same set of axes, sketch the graphs of equations $y = \cos 2x$ and $y = \tan x$ in the domain $-\pi \le x \le \pi$.
(b) What is a line of symmetry of the graph of $y = \cos 2x$ as sketched in answer to part **(a)**?

ANSWERS TO SELECTED EXERCISES: CHAPTER 7

Section 7.1
1. (a) 0.5000   **(c)** 0.3881   **(e)** 0.3314
   (b) 1.0000   **(d)** 0.8258   **(f)** 1.0000
3. (a) 115°35′   **(b)** 66°22′   **(c)** 153°52′
4. (a) 29°30′   **(c)** 50°30′   **(e)** 26°40′
   (b) 68°20′   **(d)** 53°10′   **(f)** 38°40′
5. (a) $\cos R = \dfrac{12}{13}$, $\tan R = \dfrac{5}{12}$

   (b) $\sin S = \dfrac{12}{13}$, $\cos S = \dfrac{5}{13}$, $\tan S = \dfrac{12}{5}$

7. (a) $\sqrt{2}$ **(b)** $\sin A = \cos A = \dfrac{\sqrt{2}}{2}$, $\tan A = 1$

9. $\sin A = \dfrac{7}{25}$, $\cos A = \dfrac{24}{25}$, $\tan A = \dfrac{7}{24}$

11. 74.1   **13.** 48°10′   **14.** 81.3 m

15. (a) 4.1   **(b)** 5.9

Section 7.2

1. (a) $\dfrac{\pi}{2}$ **(c)** $\dfrac{11\pi}{6}$ **(e)** $-\dfrac{5\pi}{4}$

   (b) $\dfrac{5\pi}{9}$ **(d)** $\dfrac{11\pi}{10}$ **(f)** $\dfrac{2\pi}{3}$

3. $\dfrac{2}{3}$

5. (a) 0.8391   **(b)** 1.0485   **(c)** 1.0619
6. (a) 71°40′   **(c)** 68°0′   **(e)** 50°30′
   (b) 30°30′   **(d)** 17°40′   **(f)** 72°40′
7. 6

9. $\sin\theta = \dfrac{\sqrt{5}}{3}$, $\cos\theta = -\dfrac{2}{3}$, $\tan\theta = -\dfrac{\sqrt{5}}{2}$

   $\csc\theta = \dfrac{3}{\sqrt{5}}$, $\sec\theta = -\dfrac{3}{2}$, $\cot\theta = -\dfrac{2}{\sqrt{5}}$

11. $\cos\theta = -\dfrac{15}{17}$ **13.** 9.42   **14.** 18   **15.** 10   **16.** 0.8π

Section 7.3

1. $-\cos 75°$ **10.** -1 **19.** 1   **28.** 2

2. $\tan 70°$ **11.** $-\dfrac{\sqrt{3}}{2}$ **20.** $-\dfrac{2}{\sqrt{3}}$ **29.** -2

3. $-\sin 70°$ **12.** $\dfrac{1}{2}$ **21.** 45, 135   **30.** 3

4. $\csc 50°$ **13.** -2 **22.** 210, 330   **31.** 90, 270
5. $-\sin 60°$ **14.** 1   **23.** 135, 315   **32.** 0, 180

6. $-\cos 30°$ **15.** $-\dfrac{\sqrt{2}}{2}$ **24.** 150, 210   **33.** 270

7. $\tan 45°$ **16.** $-\dfrac{1}{\sqrt{3}}$ **25.** 30, 210   **34.** 135, 315

8. $-\sec 35°$ **17.** $-\dfrac{\sqrt{3}}{2}$ **26.** 60, 300   **35.** 60, 300

9. $\dfrac{\sqrt{3}}{2}$ **18.** $-\dfrac{\sqrt{3}}{2}$ **27.** 1

Section 7.4

1. (a) $y = 3\cos x$ **(b)** $y = 3\cos 2x$ **(c)** $y = 3\cos\dfrac{1}{2}x$

2. (a) $y = 2 \sin x$ **(b)** $y = 2 \sin 2x$ **(c)** $y = 2 \sin \dfrac{1}{2}x$

3. (a) $y = 3$ **5.** -2 **8.** (3)   **11. (b)** 2

    (b) $y = 2$ **6.** 3   **9.** (2)   **12. (b)** *(1)* 3   *(2)* 4

4. (a) π **(b)** $\dfrac{\pi}{2}$ **7.** II   **10.** (3)   **13. (b)** 3

15.

Section 7.5
1. (2)   **2. (b)** 3   **3. (b)** 1

Section 7.6
1. (a) $90.0°$ **10. (a)** $135.3°$ **19.** $90°$ **28.** $\sqrt{7}$
    (b) $90° \, 0'$ **(b)** $135° \, 20'$

2. (a) $60.0°$ **11.** $-90°$ **20.** 0 **29.** $-\dfrac{7}{24}$

    (b) $60° \, 0'$

3. (a) $90.0°$ **12.** $30°$ **21.** $\dfrac{\sqrt{2}}{2}$ **30.** $\dfrac{\sqrt{5}}{3}$

    (b) $90° \, 0'$

4. (a) $-30.0°$ **13.** $-60°$ **22.** 1 **31.** $\dfrac{1}{\sqrt{10}}$

    (b) $-30° \, 0'$

5. (a) $45.9°$ **14.** $120°$ **23.** $-\dfrac{1}{2}$ **32.** $\dfrac{13}{12}$

    (b) $45° \, 0'$

6. (a) $25.2°$ **15.** $45°$ **24.** $\dfrac{3}{\sqrt{10}}$ **33.** $\dfrac{2}{\sqrt{3}}$

    (b) $25° \, 10'$

7. (a) $-270.0°$ **16.** $60°$ **25.** $\dfrac{12}{13}$ **34.** $60°$

    (b) $-270° \, 0'$

8. (a) $-30.0°$ **17.** $90°$ **26.** $\dfrac{4}{3}$ **35.** 1

    (b) $-30.\,0'$

9. (a) $-72.2°$ **18.** $90°$ **27.** 2 **36.** $-\dfrac{1}{2}$

    (b) $-72° \, 10'$

Regents Tune-Up: Chapter 7

1. 180   **4.** $\dfrac{\sqrt{3}}{4}$ **7.** (1)   **10.** (2)   **13.** (4)

2. $-\dfrac{4}{3}$ **5.** (2)   **8.** (2)   **11.** (1)   **14.** (3)

3. $-\sin 50°$ **6.** 4   **9.** (1)   **12.** (2)   **15.** (1)

16.

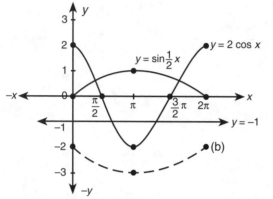

17. (a)   **(b)** y-axis

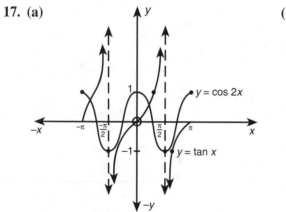

CHAPTER 8

TRIGONOMETRIC IDENTITIES, EQUATIONS, AND FORMULAS

8.1 USING QUOTIENT AND PYTHAGOREAN IDENTITIES

KEY IDEAS

An **identity** is an equation that is true for all possible replacements of the variable. Thus

$$x + 3 + x = 2x + 3$$

is an example of an identity. Several important trigonometric identities can be derived by working with a circle whose radius is 1.

Some trigonometric identities involve *squares* of trigonometric functions. The square of the sine function, for example, is written as $\sin^2 \theta$. Thus

$$(\sin \theta)(\sin \theta) = \sin^2 \theta.$$

Do not confuse $\sin^2 \theta$ with $\sin \theta^2$. The expression $\sin^2 \theta$ means that the function value is being squared, while $\sin \theta^2$ means that the measure of the angle is being squared.

Quotient Identities

Since tangent is defined as the quotient of y divided by x,

$$\tan \theta = \frac{y}{x} = \frac{\sin \theta}{\cos \theta} \quad \text{and} \quad \cot \theta = \frac{x}{y} = \frac{\cos \theta}{\sin \theta}.$$

These relationships are known as the **quotient identities.**

MATH FACTS

QUOTIENT IDENTITIES

- $\tan \theta = \dfrac{\sin \theta}{\cos \theta}$, provided that $\cos \theta \neq 0$.

- $\cot \theta = \dfrac{\cos \theta}{\sin \theta}$, provided that $\sin \theta \neq 0$.

Unit Circles

A **unit circle** is a circle whose radius is 1 and whose center is at the origin. Trigonometric functions of an angle have been defined in terms of *any* point $P(x, y)$ on the terminal side of the angle. If $P(x, y)$ is chosen so that it is the point at which the terminal side intersects the unit circle, as shown in Figure 8.1, then $OP = r = 1$, and

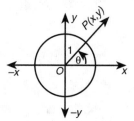

$$\cos \theta = \frac{x}{r} = \frac{x}{1} = x \quad \text{and} \quad \sin \theta = \frac{y}{r} = \frac{y}{1} = y.$$

Figure 8.1 A Unit Circle

Thus each point on the unit circle has $\cos \theta$ as its x-coordinate, and $\sin \theta$ as its y-coordinate, that is,

$$P(x, y) = P(\cos \theta, \sin \theta).$$

In a unit circle, $x^2 + y^2 = 1$. Replacing x by $\cos \theta$ and y by $\sin \theta$ produces the following **Pythagorean trigonometric identity**:

$$(\cos \theta)^2 + (\sin \theta)^2 = 1 \quad \text{or} \quad \sin^2 \theta + \cos^2 \theta = 1.$$

Dividing each term of $\sin^2 \theta + \cos^2 \theta = 1$ by either $\sin^2 \theta$ or $\cos^2 \theta$ produces two additional Pythagorean identities that are summarized in the table that follows. These identities are true for *any* angle θ. The Pythagorean identities are so important, and are needed so frequently, that they should be memorized.

PYTHAGOREAN IDENTITIES	EQUIVALENT FORMS
$\sin^2 \theta + \cos^2 \theta = 1$	$\sin^2 \theta = 1 - \cos^2 \theta$ or $\cos^2 \theta = 1 - \sin^2 \theta$
$\tan^2 \theta + 1 = \sec^2 \theta$	$\tan^2 \theta = \sec^2 \theta - 1$ or $1 = \sec^2 \theta - \tan^2 \theta$
$\cot^2 \theta + 1 = \csc^2 \theta$	$\cot^2 \theta = \csc^2 \theta - 1$ or $1 = \csc^2 \theta - \cot^2 \theta$

Examples

1. A point moves along a unit circle O in the counterclockwise direction from $A(1,0)$ to $P(-0.6, -0.8)$. Find, to the *nearest degree*, the angle of rotation.

Solution: Let θ represent the angle of rotation.

- Since $P(-0.6, -0.8)$ is on unit circle O, $x = -0.6$, $y = -0.8$, and $r = 1$. Hence θ terminates in Quadrant III, where both x and y are negative.

- Let $\cos \theta = \dfrac{x}{r} = -\dfrac{0.6}{1} = -0.6$.

  Use your scientific calculator to find that $\theta_R \approx 53°$.

- Hence, $\theta = 180° + 53° = \mathbf{233°}$.

2. If $\cos x = -\dfrac{5}{13}$ and $\angle x$ is in Quadrant III:

   (a) Use a Pythagorean identity to find $\sin x$.
   (b) Use a quotient identity to find $\tan x$.

Solutions: **(a)** Use a form of the Pythagorean identity $\sin^2 x + \cos^2 x = 1$:

$$\sin^2 x = 1 - \cos^2 x = 1 - \left(-\frac{5}{13}\right)^2 = 1 - \frac{25}{169} = \frac{169 - 25}{169} = \frac{144}{169}.$$

If $\sin^2 x = \dfrac{144}{169}$, then $x = \pm\sqrt{\dfrac{144}{169}} = \pm\dfrac{12}{13}$. Since sine is negative in Quadrant III,

$$\sin x = -\frac{12}{13}.$$

(b) Using the quotient identity with $\sin x = -\dfrac{12}{13}$ and $\cos x = -\dfrac{5}{13}$ gives

$$\tan x = \frac{\sin x}{\cos x} = \left(-\frac{12}{13}\right) \div \left(-\frac{5}{13}\right) = \left(-\frac{12}{13}\right) \cdot \left(-\frac{13}{5}\right) = \frac{12}{5}.$$

3. Express $\cos A(\sec A - \cos A)$ in simplest form in terms of $\cos A$ or $\sin A$.

Solution: Use the distributive property and then simplify using reciprocal and Pythagorean identities.

$$\cos A (\sec A - \cos A) = \cos A \sec A \quad - (\cos A)(\cos A)$$

$$= \cos A \left(\frac{1}{\cos A} \right) - \cos^2 A$$

$$= 1 - \cos^2 A$$

Since $\sin^2 A + \cos^2 A = 1$,
$\sin^2 A = 1 - \cos^2 A$:
$$= \mathbf{\sin^2 A}$$

4. Express as a single trigonometric function:

$$\frac{\sin^2 A}{(\sec A - 1)(\sec A + 1)}.$$

Solution: Multiply the binomials in the denominator of the given fraction:

$$\frac{\sin^2 A}{(\sec A - 1)(\sec A + 1)} = \frac{\sin^2 A}{\sec^2 A - 1}$$

Since $1 + \tan^2 A = \sec^2 A$,
$\sec^2 A - 1 = \tan^2 A$:
$$= \frac{\sin^2 A}{\tan^2 A}$$

Use the quotient identity:
$$= \sin^2 A \div \left(\frac{\sin^2 A}{\cos^2 A} \right)$$

Invert and multiply:
$$= \sin^2 A \cdot \frac{\cos^2 A}{\sin^2 A}$$

Simplify:
$$= \mathbf{\cos^2 A}$$

Exercise Set 8.1

1. If $\sin A = -\dfrac{3}{5}$ and $\angle A$ is in Quadrant IV, find $\cos A$.

2. If $\tan A = \dfrac{12}{5}$ and $\angle A$ is in Quadrant III, find $\sec A$.

3. The expression $(1 - \cos x)(1 + \cos x)$ is equivalent to

(1) $\sin x$ (2) $-\sin x$ (3) $\sin^2 x$ (4) $-\sin^2 x$

4. The expression $\dfrac{\tan x}{\cot x}+1$ is equivalent to

    (1) $\sin^2 x$ (2) $\cos^2 x$ (3) $\csc^2 x$ (4) $\sec^2 x$

5. The expression $\dfrac{2\sin^2 x+2\cos^2 x}{\sec x+\sec x}$ is equivalent to

    (1) $\cos x$ (2) $\sin x$ (3) $\sec x$ (4) 1

6–11. Use one or more identities to simplify each expression.

6. $\cot^2 A(1-\cos^2 A)$ **9.** $\sin x(\csc^2 x-1)$

7. $\cos A(1+\tan^2 A)$ **10.** $(1-\cos^2 A)(\csc^2 A-1)$

8. $\dfrac{1}{\cos^2 A}-1$ **11.** $(1-\sin A)(1+\sin A)\tan^2 A$

12–17. Express in simplest form in terms of sin x or cos x or both.

12. $\csc x \tan x$ **14.** $\dfrac{\tan^2 x+1}{\cot^2 x+1}$ **16.** $\dfrac{\sec x-\cos x}{\tan x}$

13. $\dfrac{\cot x}{\csc x}$ **15.** $\dfrac{\tan x+\sec x}{1+\sin x}$ **17.** $\dfrac{2\cot x}{1+\cot^2 x}$

18. If $\log \sin x = a$ and $\log \cos x = b$, write in terms of a or b, or both a and b:

    (a) $\log \sin^2 x$ **(c)** $\log \sec x$

    (b) $\log (\sec^2 A - 1)$ **(d)** $\log (\cot^2 x + 1)$

19. If $\cos x \neq -1$, the fraction $\dfrac{\sin^2 x}{1+\cos x}$ is equivalent to:

    (1) $1-\cos x$ (2) $-\cos x$ (3) $1+\cos x$ (4) $\cos x - 1$

20. Show that the expression $\dfrac{\tan\theta\csc^2\theta}{1+\tan^2\theta}$ can be reduced to $\cot\theta$.

21. Show that the expression $\dfrac{\sin\theta\tan\theta+\cos\theta}{\cos\theta}$ can be reduced to $\sec^2\theta$.

22. Show that the expression $\dfrac{\sec x}{\cot x+\tan x}$ can be reduced to $\sin x$.

23. Show that, if x is an acute angle, $\tan x = \dfrac{\sin x}{\sqrt{1-\sin^2 x}}$

199

8.2 SOLVING TRIGONOMETRIC EQUATIONS

$$\wedge \atop \text{KEY IDEAS} \atop \diagup \diagdown$$

Solving a trigonometric equation means finding the measures of the angles in a *specified* interval that make the trigonometric equation a true statement. For example, suppose that

$$\sin x = \frac{1}{2} \quad \text{and} \quad 0° \le x \le 360°.$$

Although $\sin 390° = \frac{1}{2}$, $x = 390°$ is *not* a solution since 390° does not lie between 0° and 360°.

When solving a trigonometric equation, treat the trigonometric function of the unknown angle as the variable and solve for it first.

Solving Trigonometric Equations

To solve a trigonometric equation, use familiar algebraic methods to isolate the trigonometric function. Then find the angle.

Example

1. Solve for x $(0° \le x \le 360°)$: $2 \sin x + 1 = 0$.

Solution: $\qquad\qquad\qquad\qquad 2 \sin x + 1 = 0$

Solve for $\sin x$: $\qquad\qquad\qquad 2 \sin x = -1$

Since reference angle = 30°: $\qquad\quad \sin x = -\frac{1}{2}$

Sine is negative in $\qquad\qquad$ *Q$_{III}$: $x_1 = 210°$

Quadrants III and IV: $\qquad\qquad$ Q$_{IV}$: $x_2 = 330°$

In the interval $0° \le x < 360°$, the values of x that satisfy the given equation are **210°** and **330°**.

Note: The Roman numeral subscript that is one-half line down from "Q" indicates the quadrant in which the terminal side of the angle to its right lies. The roots are numbered sequentially as $x_1, x_2, x_3, \ldots$.

2. Solve for x ($0° \le x < 360°$): $2 \sin^2 x + \sin x = 0$.

Solution: Solve for $\sin x$: $2 \sin^2 x - \sin x = 0$.

Factor $\sin x$: $\sin x (2 \sin x - 1) = 0$

$\sin x = 0$ or $2 \sin x - 1 = 0$

$x_1 = 0°$ $\sin x = \dfrac{1}{2}$

$x_2 = 180°$ $Q_I: x_3 = 30°$

$Q_{II}: x_4 = 150°$

Hence, in the interval $0° \le x < 360°$, the values of x that satisfy the given equation are **0°, 30°, 150°**, and **180°**.

3. Solve for x ($0° \le x < 180°$): $4 \cos^2 x - 3 = 0$.

Solution: $4 \cos^2 x - 3 = 0$

$\cos^2 x = \dfrac{3}{4}$

Take the square root of each side: $\cos x = \pm\sqrt{\dfrac{3}{4}} = \pm\dfrac{\sqrt{3}}{2}$

$\cos x = \dfrac{\sqrt{3}}{2}$ or $\cos x = -\dfrac{\sqrt{3}}{2}$

$Q_I: x_1 = 30°$ $Q_{II}: x_3 = 150°$

$Q_{IV}: x_2 = 330°$ $Q_{III}: x_4 = 210°$

In the interval $0° \le x < 180°$, the values of x that satisfy the equation are **30°** and **150°**.

4. Solve for x to the *nearest tenth of a degree* ($0° \le x < 360°$): $\tan^2 x = \tan x + 2$.

Solution: Write the quadratic equation in standard form and factor:

$$\tan^2 x - \tan x - 2 = 0$$
$$(\tan x + 1)(\tan x - 2) = 0$$

$\tan x + 1 = 0$ or $\tan x - 2 = 0$

$\tan x = -1$ $\tan x = 2$

$Q_{II}: x_1 = 135°$ $x = \tan^{-1} 2$

$Q_{IV}: x_2 = 315°$ Using a scientific calculator, find x to the *nearest tenth of a degree:* 63.4°.

$Q_I: x_3 = 63.4°$

$Q_{III}: x_4: = 243.4°$

In the given interval $0° \le x < 360°$, the values of x to the *nearest tenth of a degree* that satisfy the given equation are **63.4°, 135°, 243.4°**, and **315°**.

Equations Requiring Substitutions

A trigonometric equation may include two different trigonometric functions. A reciprocal or Pythagorean trigonometric identity may permit one of these functions to be replaced so that an equivalent equation results that contains powers of the same trigonometric function.

Examples

5. For all values in the interval $0° \leq x < 360°$ for which the functions are defined, solve $2 \cos x - 3 = 2 \sec x$.

Solution: The equation $2 \cos x - 3 = 2 \sec x$ includes the cosine and secant functions. Since these are reciprocal functions, $\sec x$ may be replaced by $\dfrac{1}{\cos x}$ ($\cos x \neq 0$):

$$2 \cos x - 3 = 2 \sec x$$

$$= 2\left(\frac{1}{\cos x}\right)$$

To eliminate the fractional term, multiply each term of the equation by $\cos x$:

$$(\cos x)(2 \cos x) - 3 \cos x = (\cos x)\left(\frac{2}{\cos x}\right)$$

$$2 \cos^2 x - 3 \cos x = 2$$
$$2 \cos^2 x - 3 \cos x - 2 = 0$$
$$(2 \cos x + 1)(\cos x - 2) = 0$$

$$2 \cos x + 1 = 0 \qquad \text{or} \qquad \cos x - 2 = 0$$
$$\cos x = -\frac{1}{2} \qquad\qquad\qquad \cos x = 2$$

$\text{Q}_{\text{II}}: x_2 = 120°$ Since the maximum value of the cosine
$\text{Q}_{\text{III}}: x_2 = 240°$ function is 1, reject this root.

In the interval $0° \leq x < 360°$, the values of x that satisfy the given equation are **120°** and **240°**.

6. Solve for x to the *nearest tenth of a degree* or to the *nearest 10 minutes* ($0° \leq x < 360°$):

$$3 \cos^2 x + 5 \sin x = 4.$$

Solution: The sine and cosine functions are related by the identity $\sin^2 x + \cos^2 x = 1$. When $\cos^2 x$ is replaced with $1 - \sin^2 x$ in the original equation, an equivalent equation results that contains only the sine function.

$$3\cos^2 x + 5\sin x = 4$$

Replace $\cos^2 x$: $\qquad\qquad 3(1 - \sin^2 x) + 5\sin^2 x = 4$

Use the distributive law: $\qquad 3 - 3\sin^2 x + 5\sin x = 4$

Simplify: $\qquad\qquad\qquad -3\sin^2 x + 5\sin x - 1 = 0$

Multiply by -1: $\qquad\qquad 3\sin^2 x - 5\sin x + 1 = 0$

Since the quadratic equation cannot be factored, use the quadratic formula to solve for $\sin x$, where $a = 3$, $b = -5$, and $c = 1$:

$$\sin x = \frac{-b \pm \sqrt{b^2 - 4ac}}{2a}$$

$$= \frac{-(-5) \pm \sqrt{(-5)^2 - 4(3)(1)}}{2 \cdot 3}$$

$$= \frac{+5 \pm \sqrt{25 - 12}}{6}$$

$$= \frac{5 \pm \sqrt{13}}{6}$$

Since $\sqrt{13} \approx 3.60555$: $\qquad \sin x \approx \dfrac{5 \pm 3.60555}{6}$

$$\sin x \approx \frac{5 - 3.60555}{6} \quad \text{or} \quad \sin x \approx \frac{5 + 3.60555}{6}$$

$$\approx 0.2324 \qquad\qquad\qquad \approx 1.39343$$

$$x \approx \sin^{-1} 0.2324$$

Q_I: $x_1 \approx 13°30'$

Q_{II}: $x_2 \approx 166°30'$

Since the maximum value of the sine function is 1, reject this root.

In the interval $0° \le x < 360°$, the values of x, correct to the *nearest 10 minutes,* that satisfy the given equation are **13°30′** and **166°30′**, or to the *nearest tenth of a degree*, 13.5° and 166.5°.

Exercise Set 8.2

1–12. Solve for x, *where* $0° \le x < 360°$.

1. $2\tan x + 2 = 0$

2. $4\sin x + 5 = 5\sin x + 4$

3. $|2\sin x - 1| = 2$

4. $2\sin^2 x - 1 = 0$

5. $|\tan x - 1| = 2$

6. $\tan^2 x - 3 = 0$

7. $4\cos^2 x - 1 = 0$

8. $4\sin^2 x - 3 = 0$

9. $2\sin^2 x + \sin x = 1$

10. $3\tan^2 x + \sqrt{3}\tan x = 0$

11. $\cos^2 x + 2 \cos x = 0$ **12.** $\sec^2 x = \sec x + 2$

13–18. Solve for x *($0° \le$ x $< 360°$) by making an appropriate substitution, using either a reciprocal or a Pythagorean identity.*

13. $2 \csc x + 3 = 2 \sin x$ **16.** $2 \sin^2 x - 3 \cos x = 3$

14. $\csc^2 x + \cot x = 1$ **17.** $2 \cos^2 x = 3(\sin x + 1)$

15. $1 + 2 \cos x = \sec x$ **18.** $2 \sec^2 x + \tan x = 5$

19–28. Solve for x *to the nearest tenth of a degree or the nearest 10 minutes, where $0° \le$ x $< 360°$.*

19. $3 \tan x + 4 = 0$ **24.** $3 \cos x + 1 = \sec x$

20. $5 \cos^2 x + \cos x = 0$ **25.** $2(\tan x - \cot x) + 3 = 0$

21. $3 \sin^2 x + 5 \sin x = 2$ **26.** $2 \cos^2 x + 3 \cos x = 3$

22. $2 \cot^2 x + 5 \cot x + 3 = 0$ **27.** $\cos^2 x = 4 \sin x + 2$

23. $\sin x + 2 = 3 \sin^2 x$ **28.** $\csc^2 x = 7 \csc x + 8$

29. Find, correct to the *nearest tenth of a degree* or the *nearest 10 minutes,* all values of x in the interval $0° \le x < 360$ that satisfy the equation

$$2 \tan^2 x - 5 \tan x = 2.$$

30. Find, correct to the *nearest tenth of a degree,* or the *nearest 10 minutes,* all values of x in the interval $0° \le x < 360°$ that satisfy the equation

$$\frac{3 \sin x}{6 \sin x + 1} = \frac{\csc x}{3}.$$

8.3 APPLYING THE SUM AND DIFFERENCE IDENTITIES

KEY IDEAS

Although $\sin (A + B) \ne \sin A + \sin B$, there are two identities that allow trigonometric functions of the sum or difference of two angles to be expressed in terms of combinations of trigonometric functions of the individual angles.

Formulas for Sin ($A \pm B$)

The sine of the sum or the difference of two angles can be expressed in terms of the sines and cosines of the individual angles.

MATH FACTS

SINE OF SUM OR DIFFERENCE

$$\sin (A + B) = \sin A \cos B + \cos A \sin B$$
$$\sin (A - B) = \sin A \cos B - \cos A \sin B$$

Examples

1. What is the exact value of $\sin 17° \cos 13° + \cos 17° \sin 13°$?

Solution: The given expression takes the form

$$\sin (A + B) = \sin A \cos B + \cos A \sin B,$$

where $A = 17°$ and $B = 13°$. Hence

$$\sin 17° \cos 13° + \cos 17° \sin 13° = \sin (17° + 13°)$$
$$= \sin 30°$$
$$= \frac{1}{2}.$$

2. Given: $\sin A = \dfrac{4}{5}$ and $\cos B = \dfrac{12}{13}$. If $\angle A$ is obtuse and $\angle B$ is acute, find the value of $\sin (A + B)$.

Solution: Before the identity for $\sin (A + B)$ can be used, the values of $\cos A$ and $\sin B$ must be determined.

- Find $\cos A$ by locating the reference triangle. Since

$$\sin A = \frac{4}{5} = \frac{y}{r}$$

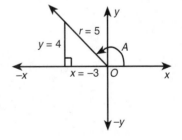

and $\angle A$ terminates in Quadrant II, let $y = 4$ and $r = 5$. The reference triangle is a 3-4-5 right triangle, where $x = -3$. Hence

205

$$\cos A = \frac{x}{r} = -\frac{3}{5}.$$

- Find sin B by locating the reference triangle.

$$\cos B = \frac{y}{r} = \frac{5}{13}.$$

- Evaluate the identity for sin $(A + B)$.

$$\sin(A + B) = \sin A \cos B + \cos A \sin B$$

$$= \left(\frac{4}{5}\right)\left(\frac{12}{13}\right) + \left(-\frac{3}{5}\right)\left(\frac{5}{13}\right)$$

$$= \frac{48}{65} \quad - \frac{15}{65}$$

$$= \frac{33}{65}$$

3. If cos $x = a$, express the value of sin $(270° - x)$ in terms of a.

Solution: Use the identity

$$\sin(A - B) = \sin A \cos B - \cos A \sin B,$$

where $A = 270°$ and $B = x$.

$$\sin(270 - x)° = \sin 270° \cos x - \cos 270° \sin x$$
$$= (-1)(a) - 0 \cdot \sin x$$
$$= -a$$

Formulas for Cos (*A* ± *B*)

The cosine of the sum or the difference of two angles can be expressed in terms of the cosines and sines of the individual angles.

MATH FACTS

COSINE OF SUM OR DIFFERENCE

cos (*A* + *B*) = cos *A* cos *B* − sin *A* sin *B*
cos (*A* − *B*) = cos *A* cos *B* + sin *A* sin *B*

Examples

4. Using functions of 60° and 45°, express the value of cos 15° in radical form.

Solution: Since 60° – 45° = 15°, use the identity

$$\cos (A - B) \cos A \cos B + \sin A \sin B,$$

where $A = 60°$ and $B = 45°$.

$$\cos(60 - 45)° = \cos 60° \cos 45° + \sin 60° \cos 45°$$

$$= \left(\frac{1}{2}\right)\left(\frac{\sqrt{2}}{2}\right) \quad + \left(\frac{\sqrt{3}}{2}\right)\left(\frac{\sqrt{2}}{2}\right)$$

$$= \frac{\sqrt{2}}{4} \qquad + \frac{\sqrt{6}}{4}$$

$$= \frac{\sqrt{2} + \sqrt{6}}{4}$$

5. If $\angle\theta$ is in Quadrant II and $\tan \theta = -\dfrac{4}{3}$, find the value of $\cos (180 + \theta)°$.

Solution: Use the identity

$$\cos (A + B) = \cos A \cos B - \sin A \sin B,$$

where $A = 180°$ and $B = \theta$. The values of $\cos \theta$ and $\sin \theta$ must be determined. Since

$$\tan \theta = -\frac{4}{3} = \frac{y}{x}$$

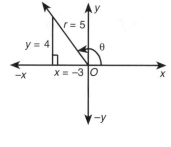

and $\angle\theta$ is in Quadrant II, let $x = -3$ and $y = 4$. The reference triangle is a 3-4-5 right triangle, where $r = 5$. Hence

$$\cos \theta = \frac{x}{r} = -\frac{3}{5} \quad \text{and} \quad \sin \theta = \frac{y}{r} = \frac{4}{5}.$$

Thus

$$\cos (180 + \theta)° = \cos 180° \cos \theta - \sin 180° \sin \theta$$

$$= (-1)\left(-\frac{3}{5}\right) - 0 \cdot \left(\frac{4}{5}\right)$$

$$= \frac{3}{5}$$

Formulas for Tan $(A \pm B)$

The tangent of the sum or the difference of two angles can be expressed in terms of the tangents of the individual angles.

MATH FACTS

TANGENT OF SUM OR DIFFERENCE

$$\tan(A + B) = \frac{\tan A + \tan B}{1 - \tan A \tan B} \quad \text{and} \quad \tan(A - B) = \frac{\tan A - \tan B}{1 + \tan A \tan B}$$

Examples

6. If $\tan x = a$, express $\tan(-x)$ in terms of a.

Solution: Use the identity for $\tan(A - B)$, where $A = 0°$ and $B = x$.

$$\tan(A - B) = \frac{\tan A - \tan B}{1 + \tan A \tan B}$$

$$\tan(0 - x) = \frac{\tan 0 - \tan x}{1 + \tan 0 \tan x}$$

$$\tan(-x) = \frac{0 - \tan x}{1 + 0} = -\tan x = -a$$

7. If $\tan \theta = \frac{2}{3}$, find the value of $\tan(\theta + 45°)$.

Solution: Evaluate

$$\tan(A + B) = \frac{\tan A + \tan B}{1 - \tan A \tan B},$$

where $\tan A = \tan \theta = \frac{2}{3}$ and $\tan B = \tan 45° = 1$.

$$\tan(\theta + 45°) = \frac{\frac{2}{3} + 1}{1 - \left(\frac{2}{3}\right)(1)} = \frac{\frac{5}{3}}{\frac{1}{3}} = 5$$

Formulas for Sin $(-x)$, Cos $(-x)$, and Tan $(-x)$

In Example 6, it was shown that $\tan(-x) = -\tan x$. In a similar manner, reduction formulas for $\sin(-x)$ and $\cos(-x)$ can also be derived (see Exercise 6 below).

FUNCTIONS OF NEGATIVE ANGLES

$$\sin (-x) = -\sin x \qquad \cos (-x) = \cos x \qquad \tan (-x) = -\tan x$$

Exercise Set 8.3

1–4. Use an appropriate identity to find the exact value.

1. $\cos 34° \cos 26° - \sin 34° \sin 26°$

3. $\sin 75° \cos 15° + \cos 75° \sin 15°$

2. $\sin 50° \cos 5° - \cos 50° \sin 5°$

4. $\cos 80° \cos 20° + \sin 80° \sin 20°$

5. If $\tan x = \dfrac{1}{2}$ and $\tan y = \dfrac{2}{3}$, find the value of $\tan (x - y)$.

6. (a) Starting with the identity for $\sin (A - B)$, prove that $\sin (-x) = -\sin x$.
   (b) Starting with the identity for $\cos (A - B)$, prove that $\cos (-x) = \cos x$.

7. If $\sin A = \dfrac{5}{13}$, $\cos B = -\dfrac{4}{5}$, $\angle A$ is acute, and $\angle B$ is obtuse, find the numerical value of:
   (a) $\sin (A - B)$ **(c)** $\tan (45° - B)$
   (b) $\cos (A + B)$ **(d)** $\sec (A - B)$

8. If $\tan A = \dfrac{8}{15}$ and $\angle A$ terminates in Quadrant III, find the numerical value of:
   (a) $\sin (90° + A)$ **(c)** $\tan (A + 135°)$
   (b) $\cos (180° - A)$ **(d)** $\csc (270 - A)$

9. Using the identity for $\tan (A - B)$, show that $\tan 15° = 2 - \sqrt{3}$.

10. If $\tan (x + y) = 4$ and $\tan y = 2$, find the numerical value of $\tan x$.

11. If $\tan x = a$, express $\tan (45° - x)$ in terms of a.

12. Using functions of 45° and 30°, express in radical form the value of:
   (a) $\sin 15°$ **(b)** $\cos 75°$ **(c)** $\tan 15°$ **(d)** $\tan 75°$

13. The expression $\cos (\pi + x) + \sin (\pi + x)$ equals:
   (1) $\cos x + \sin x$ (3) $-\cos x + \sin x$
   (2) $\cos x - \sin x$ (4) $-\cos x - \sin x$

14. The expression $\cos (270° + x)$ is equivalent to:
   (1) $\sin x$ (2) $-\sin x$ (3) $\cos x$ (4) $-\cos x$

15. The expression $\dfrac{\sin{(90° + x)}}{\sin{(-x)}}$ can be reduced to:

(1) –1   (2) 1   (3) –cot x (4) cot x

16. If $\sin A = \dfrac{3}{5}$, $\sin B = \dfrac{5}{13}$, and angles A and B are acute angles, what is the value of $\cos{(A - B)}$?

(1) $-\dfrac{12}{65}$ (2) $\dfrac{16}{65}$ (3) $\dfrac{33}{65}$ (4) $\dfrac{63}{65}$

8.4 APPLYING THE DOUBLE-ANGLE IDENTITIES

KEY IDEAS

Letting $A = B$ in the identities for sin $(A + B)$, cos $(A + B)$, and tan $(A + B)$ produces the double-angle identities for sin $2A$, cos $2A$, and tan $2A$.

Formula for Sin 2A

Replacing $\angle B$ with $\angle A$ in the identity for sin $(A + B)$ yields the identity for sin $2A$.

MATH FACTS

SINE OF DOUBLE ANGLE

sin 2A = 2 sin A cos A

Examples

1. Find the value of sin $2A$ if $\sin A = \dfrac{3}{5}$ and $\angle A$ is obtuse.

Solution: Since sin $2A = 2$ sin A cos A, the value of cos A is needed. Angle A terminates in Quadrant II with $\sin A = \dfrac{3}{5} = \dfrac{y}{r}$. The reference triangle is a 3-4-5 right triangle with $x = -4$. Hence $\cos A = -\dfrac{4}{5}$.

$$\sin 2A = 2 \sin A \cos A$$
$$= 2\left(\frac{3}{5}\right)\left(-\frac{4}{5}\right) = -\frac{24}{25}$$

2. If $\theta = \text{Arc tan } \dfrac{2}{\sqrt{5}}$, find the value of sin 2θ.

Solution: To evaluate the identity for sin 2θ, first find the values of sin θ and cos θ.

- Since $\theta = \text{Arc tan } \dfrac{2}{\sqrt{5}}$, then

$$\tan \theta = \frac{2}{\sqrt{5}} = \frac{y}{x}$$

and θ is in Quadrant I. Find the value of r by letting $x = \sqrt{5}$ and $y = 2$ in the Pythagorean relationship $x^2 + y^2 = r^2$:

$$(\sqrt{5})^2 + 2^2 = r^2$$
$$5 + 4 = r^2$$

- Since $r^2 = 9$, $r = 3$. Hence

$$\sin \theta = \frac{y}{r} = \frac{2}{3}$$

and

$$\cos \theta = \frac{x}{r} = \frac{\sqrt{5}}{3}.$$

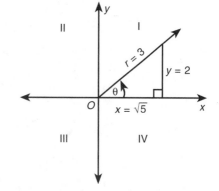

- Use the double-angle identity:

$$\sin 2\theta = 2 \sin \theta \cos \theta = 2\left(\frac{2}{3}\right)\left(\frac{\sqrt{5}}{3}\right) = \frac{4\sqrt{5}}{9}.$$

Formula for Cos 2A

Replacing $\angle B$ with $\angle A$ in the identity for cos $(A + B)$ produces the identity

$$\cos 2A = \cos^2 A - \sin^2 A.$$

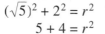
MATH FACTS

COSINE OF DOUBLE ANGLE

- **cos $2A$ = cos^2 A – sin^2 A,** or
- **cos $2A$ = 2 cos^2 A – 1,** or
- **cos $2A$ = 1 – 2 sin^2 A**

Examples

3. If $\angle x$ is in Quadrant II and $\sin x = 0.8$, find the value of $\cos 2x$.

Solution: Since the value of $\sin x$ is given, choose the form of the identity for $\cos 2x$ that involves only the sine function, that is, the third form, $\cos 2x = 1 - 2 \sin^2 x$.

$$\cos 2x = 1 - 2 \sin^2 x = 1 - 2(0.8)^2$$
$$= 1 - 2(0.64)$$
$$= 1 - 1.28$$
$$= -0.28$$

4. Show that the expression $\dfrac{\cos 2x}{\cos x + \sin x}$ is equivalent to $\cos x - \sin x$.

Solution: Choose the form of the identity for $\cos 2x$ that involves cosine and sine.

$$\frac{\cos 2x}{\cos x + \sin x} = \frac{\cos^2 x - \sin^2 x}{\cos x + \sin x}$$

Factor: $$= \frac{\overset{1}{\cancel{(\cos x + \sin x)}}(\cos x - \sin x)}{\cancel{\cos x + \sin x}} = \cos x - \sin x$$

5. Solve for x ($0° \le x < 360°$): $\cos 2x - \cos x = 0$.

Solution: Since the equation contains $\cos x$, choose the form of the identity for $\cos 2x$ that is expressed only in terms of the cosine function.

$$\cos 2x - \cos x = 0$$

Replace $\cos 2x$ by $2 \cos^2 x - 1$: $(2 \cos^2 x - 1) - \cos x = 0$

Put into standard form: $2 \cos^2 x - \cos x - 1 = 0$

Factor: $(2 \cos x + 1)(\cos x - 1) = 0$

$$2 \cos x + 1 = 0 \qquad \text{or} \quad \cos x - 1 = 0$$
$$\cos x = -\frac{1}{2} \qquad\qquad \cos x = 1$$
$$Q_{II}: x_1 = 120° \qquad\qquad x_3 = 0°$$
$$Q_{III}: x_2 = 240°$$

In the interval $0° \le x < 360°$, the values of x that satisfy the equation are **0°**, **120°**, and **240°**.

Formula for Tan 2A

Replacing $\angle B$ with $\angle A$ in the identity for $\tan (A + B)$ produces the identity for $\tan 2A$.

TANGENT OF DOUBLE ANGLE

$$\tan 2A = \frac{2 \tan A}{1 - \tan^2 A}$$

Example

6. If $\tan A = 0.5$, find the value of $\tan 2A$.

Solution: $\tan 2A = \dfrac{2 \tan A}{1 - \tan^2 A} = \dfrac{2(0.5)}{1 - (0.5)^2}$

$$= \frac{1}{0.75} = 1 \div \frac{3}{4} = \frac{4}{3}$$

Exercise Set 8.4

1. If $\cos x = \dfrac{15}{17}$ and $\angle x$ is acute, find the exact value.
   (a) $\sin 2x$ **(b)** $\cos 2x$ **(c)** $\tan 2x$

2. If $\tan x > 0$ and $\sin x = -\dfrac{24}{25}$, find the exact value.
   (a) $\sin 2x$ **(b)** $\cos 2x$ **(c)** $\tan 2x$

3. If $\tan x < 0$ and $\cos x = \dfrac{3}{\sqrt{13}}$, find the exact value.
   (a) $\sin 2x$ **(b)** $\cos 2x$ **(c)** $\tan 2x$

4. If $\csc \theta > 0$ and $\tan \theta = -\dfrac{1}{2}$, find the exact value.
   (a) $\sin 2\theta$ **(b)** $\cos 2x$ **(c)** $\tan 2\theta$

5. Solve for x, where $0° \le x < 360°$.
   (a) $\cos 2x - \sin x = 0$ **(d)** $\cos 2x = \sin 2x$
   (b) $\sin 2x + \sin x = 0$ **(e)** $\cos 2x - 3 \cos x = 1$
   (c) $\sin 2x + \sin (90° - x) = 0$ **(f)** $2 \sin 3x \cos 3x = 1$

6. Solve for θ to the *nearest 10 minutes,* where $0° \le \theta < 360°$.
   (a) $4 \cos 2\theta - 3 \sin \theta + 1 = 0$ **(b)** $2 \cos 2\theta = 7 \cos \theta$

7. Show that the expression $\dfrac{\sin 2A}{2 \tan A}$ is equivalent to $\cos^2 A$.

8. Show that the expression $(\cos x + \sin x)^2 - 1$ is equivalent to $\sin 2x$.

9. Show that the expression $\sin 2x \cot 2x + 1$ is equivalent to $2 \cos^2 x$.

10. Show that the expression $\sin x \sin 2x + \cos x \cos 2x$ is equivalent to $\cos x$.

11. Show that the expression $\dfrac{1 + \cos A + \cos 2A}{\sin A + \sin 2A}$ is equivalent to $\cot A$.

8.5 APPLYING THE HALF-ANGLE IDENTITIES

=== KEY IDEAS ===

Since $\angle A$ is two times as great as $\angle \dfrac{A}{2}$, the identity

$$\cos 2\theta = 1 - 2 \sin^2 \theta$$

is true when 2θ is replaced by A and θ is replaced by $\dfrac{A}{2}$. Solving for $\sin \dfrac{A}{2}$ gives

$$\sin \frac{A}{2} = \pm \sqrt{\frac{1 - \cos A}{2}}.$$

Similarly, the identity for $\cos \dfrac{A}{2}$ can be obtained from the identity

$$\cos 2\theta = 2 \cos^2 \theta - 1.$$

Formulas for $\text{Sin}\dfrac{A}{2}$, $\text{Cos}\dfrac{A}{2}$, and $\text{Tan}\dfrac{A}{2}$

The half-angle identities are summarized below. Notice that dividing the identity for $\sin \frac{A}{2}$ by the identity for $\cos \frac{A}{2}$ gives the identity for $\tan \frac{A}{2}$.

214

HALF-ANGLE IDENTITIES

- $\sin\dfrac{A}{2} = \pm\sqrt{\dfrac{1-\cos A}{2}}$

- $\cos\dfrac{A}{2} = \pm\sqrt{\dfrac{1+\cos A}{2}}$

- $\tan\dfrac{A}{2} = \pm\sqrt{\dfrac{1-\cos A}{1+\cos A}}$

Each half-angle formula is expressed in terms of cos A. The choice of a positive or negative sign in front of each radical depends on the quadrant in which $\angle\frac{A}{2}$ lies. For example, suppose that $0° \le A \le 360°$. If $\angle A$ lies in Quadrant IV, then $270° < A < 360°$. To determine the quadrant in which $\frac{A}{2}$ lies, divide each member of the inequality by 2. The result is

$$135° < \frac{A}{2} < 180°.$$

This means that $\frac{A}{2}$ lies in Quadrant II, so sin $\frac{A}{2}$ is positive while cos $\frac{A}{2}$ and tan $\frac{A}{2}$ are negative.

Examples

1. If $\cos x = \dfrac{7}{8}$ and $\angle x$ is acute, find the value of $\sin\dfrac{x}{2}$.

Solution: Since $\angle x$ is acute, $\dfrac{x}{2}$ is acute and $\sin\dfrac{x}{2}$ is positive. Then

$$\sin\frac{x}{2} = \sqrt{\frac{1-\cos x}{2}}$$

$$= \sqrt{\frac{1-\dfrac{7}{8}}{2}}$$

$$= \sqrt{\frac{\dfrac{1}{8}}{2}} = \sqrt{\frac{1}{16}}$$

$$= \frac{1}{4}$$

2. If tan = $\dfrac{24}{7}$ and $\angle A$ $(0° \le A < 360°)$ terminates in Quadrant III, find the value of cos $\dfrac{A}{2}$.

Solution: First find the value of cos A. To evaluate the half-angle identity for cosine:

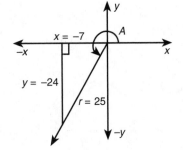

- Since $\tan A = \dfrac{24}{7} = \dfrac{y}{x}$

  and $\angle A$ terminates in Quadrant III, let $x = -7$ and $y = -24$. Since $x^2 + y^2 = r^2$,

  $$r = \sqrt{(-7)^2 + (-24)^2} = 25$$

  so

  $$\cos A = \frac{x}{r} = \frac{-7}{25}$$

- Determine the sign of cos $\dfrac{A}{2}$. If $\angle A$ is in Quadrant III, then $180° < A < 270°$, so that $90° < \dfrac{A}{2} < 135°$. Therefore, $\dfrac{A}{2}$ lies in Quadrant II, where cosine is negative.

- Use the half-angle identity:

$$\cos \frac{A}{2} = -\sqrt{\frac{1 + \cos A}{2}} = -\sqrt{\frac{1 + \left(\dfrac{-7}{25}\right)}{2}}$$

$$= -\sqrt{\frac{\dfrac{18}{25}}{2}}$$

$$= -\sqrt{\frac{1}{2}\left(\frac{18}{25}\right)}$$

$$= -\sqrt{\frac{9}{25}}$$

$$= -\frac{3}{5}$$

3. If $\cos A = -\dfrac{1}{10}$ and $\angle A$ is obtuse, find the exact value of $\tan \dfrac{A}{2}$.

Solution: If $\angle A$ is obtuse, then $90° < A < 180°$. Hence $45° < \dfrac{A}{2} < 90°$.

Since $\dfrac{A}{2}$ is acute, then tangent of $\dfrac{A}{2}$ is positive. Hence

$$\tan \frac{A}{2} = \sqrt{\frac{1 - \cos A}{1 + \cos A}} = \sqrt{\frac{1 - \left(\dfrac{-1}{10}\right)}{1 + \left(\dfrac{-1}{10}\right)}}$$

$$= \sqrt{\frac{1 + \dfrac{1}{10}}{1 - \dfrac{1}{10}}} = \sqrt{\frac{11}{10} \div \frac{9}{10}}$$

$$\tan \frac{A}{2} = \sqrt{\frac{11}{9}} = \frac{\sqrt{11}}{\sqrt{9}}$$

$$= \frac{\sqrt{11}}{3}$$

Exercise Set 8.5

1. If $\angle x$ ($0° \le x < 360°$) terminates in Quadrant IV and $\cos x = 0.28$, find each value.

    (a) $\sin \dfrac{x}{2}$ **(b)** $\cos \dfrac{x}{2}$ **(c)** $\tan \dfrac{x}{2}$ **(d)** $\sec \dfrac{x}{2}$

2. If $\angle A$ is an acute angle and $\sin A = \dfrac{\sqrt{32}}{9}$, find each value.

    (a) $\sin \dfrac{A}{2}$ **(b)** $\cos \dfrac{A}{2}$ **(c)** $\tan \dfrac{A}{2}$ **(d)** $\csc \dfrac{A}{2}$

3. Find the value of $\cos \dfrac{A}{2}$ if $\angle A$ is acute and

    (a) $\cos A = \dfrac{1}{7}$ **(b)** $\tan A = \dfrac{4}{3}$

4. Find the value of $\sin \dfrac{A}{2}$ if $\angle A$ ($0° \le A < 360°$) terminates in Quadrant III and

    (a) $\cos A = -0.62$ **(b)** $\sin A = -\dfrac{21}{29}$

5. Find the value of tan $\dfrac{\theta}{2}$ if $\angle\theta$ $(0° \le \theta < 360°)$ terminates in Quadrant IV and:

(a) $1 - \cos A = 0.2$ (b) $\cos 2A = \dfrac{-7}{9}$

6. If sin $\dfrac{\theta}{2} = 0.4$ and $\angle\theta$ is obtuse, find the value of cos θ.

8.6 PROVING TRIGONOMETRIC IDENTITIES

====================== KEY IDEAS ======================

The equation

$$2x + 1 = 9$$

is a *conditional equation* since it is true for only a particular value of the variable, namely, $x = 4$.

Since each side of the equation

$$x + 5 + x = 1 + 2x + 4$$

can be rewritten so that $2x + 5 = 2x + 5$, the original equation must be true for all possible values of x and is, therefore, an *identity*. This observation suggests a method that can be used to *prove* new trigonometric identities.

A Strategy for Proving Identities

To prove that a trigonometric equation is an identity, work on each side of the equation *separately* and show that the two sides can be made to look exactly alike. Start with the more complicated side of the equation, and express it in terms of sines and cosines, using the reciprocal and quotient identities. Working on the same side of the equation, you may need to do the following:

- Write a fraction in lowest terms, simplify a complex fraction, or combine fractions.
- Make a substitution, using another trigonometric identity.

- Perform some other algebraic operation that does not affect the other, less complicated side of the equation.

If the two sides of the equation still do not look alike, it may be helpful to transform the other side of the equation also, as illustrated in Example 1.

Examples

1. Prove that the following equation is an identity for all values of θ for which the expressions are defined:

$$\sec \theta - \cos \theta = \sin \theta \tan \theta.$$

Solution: To emphasize the importance of not performing operations that will affect both sides of the equation, draw a vertical boundary line below the equal sign. Express the left side in terms of sines and cosines, and then simplify.

$$\sec \theta - \cos \theta = \sin \theta \tan \theta$$

Use the reciprocal identity:	$\dfrac{1}{\cos \theta} - \cos \theta$	
Rewrite the second term, $\cos \theta$, as a fraction having $\cos \theta$ as its denominator:	$\dfrac{1}{\cos \theta} - \dfrac{\cos^2 \theta}{\cos \theta}$	
Add:	$\dfrac{1 - \cos^2 \theta}{\cos \theta}$	
Use the Pythagorean identity:	$\dfrac{\sin^2 \theta}{\cos \theta}$	$\sin \theta \tan \theta$
Change the right side to sines and cosines:	$\dfrac{\sin^2 \theta}{\cos \theta}$	$\sin \theta \left(\dfrac{\sin \theta}{\cos \theta} \right)$
Multiply on the right side:	$\dfrac{\sin^2 \theta}{\cos \theta}$	$= \dfrac{\sin^2 \theta}{\cos \theta}.$

Hence **$\sec \theta - \cos \theta = \sin \theta \tan \theta$ is an identity.**

2. Prove that the following equation is an identity for all values of A for which the expressions are defined:

$$\sin 2A \cot A = 1 + \cos 2A.$$

Solution: On the left side of the given equation, express $\sin 2A$ and $\cot A$ in terms of $\sin A$ and $\cos A$, using the appropriate double-angle and quotient identities. In the right side of the equation, replace $\cos 2A$ with a form of the double-angle identity that makes the two sides look alike.

$$\sin 2A \cot A = 1 + \cos 2A$$

$$(2 \sin A \cos A) \frac{\cos A}{\sin A}$$

$$2 \cos^2 A \mid 1 + \cos 2A$$

Use the identity on the right side for cos 2A:

$$2 \cos^2 A \mid 1 + (2 \cos^2 A - 1)$$
$$2 \cos^2 A = 2 \cos^2 A$$

Hence **sin 2A cot A = 1 + cos 2A is an identity.**

3. Prove that the following equation is an identity for all values of x for which the expressions are defined:

$$\frac{\sec x + \csc x}{\tan x + \cot x} = \sin x + \cos x.$$

Solution: Since the right side of the given equation is already expressed in terms of sines and cosines, work only on the left side. Write the numerator in terms of sin x and cos x, using reciprocal identities. Change the denominator to sines and cosines, using the quotient identities.

$$\frac{\sec x + \csc x}{\tan x + \cot x} = \sin x + \cos x$$

Simplify the complex fraction by multiplying its numerator and denominator by sin x cos x:

$$\frac{\dfrac{1}{\cos x} + \dfrac{1}{\sin x}}{\dfrac{\sin x}{\cos x} + \dfrac{\cos x}{\sin x}}$$

$$\frac{(\sin x \cos x)\left[\dfrac{1}{\cos x} + \dfrac{1}{\sin x}\right]}{(\sin x \cos x)\left[\dfrac{\sin x}{\cos x} + \dfrac{\cos x}{\sin x}\right]}$$

Multiply using the distributive property:

$$\frac{\sin x + \cos x}{\sin^2 x + \cos^2 x}$$

Use the Pythagorean identity:

$$\frac{\sin x + \cos x}{1}$$
$$\sin x + \cos x = \sin x + \cos x$$

Hence $\dfrac{\sec x + \csc x}{\tan x + \cot x} = \sin x + \cos x$ **is an identity.**

Exercise Set 8.6

1–15. Prove that each expression is an identity for all values of the angle(s) for which it is defined.

1. $\sin^2 A(1 + \tan^2 A) = \tan^2 A$

2. $\tan A(\csc A - \sin A) = \cos A$

3. $\dfrac{2 \cot A}{1 + \cot^2 A} = \sin 2A$

4. $\dfrac{1 + \csc A}{\sec A} = \cos A + \cot A$

5. $\tan A = \dfrac{\sin 2A}{1 + \cos 2A}$

6. $\dfrac{\sin 2A}{\sin A} - \dfrac{\cos 2A}{\cos A} = \sec A$

7. $\dfrac{2 \tan \theta}{1 + \tan^2 \theta} = \sin 2\theta$

8. $\dfrac{\cos^2 A}{\csc A - 1} = \dfrac{1 + \sin A}{\sin A}$

9. $\dfrac{(\cos x + \sin x)^2}{1 + \sin 2x} = \dfrac{\cos x \tan x}{\csc x}$

10. $\cot A - \dfrac{\cos 2A}{\sin A \cos A} = \tan A$

11. $\dfrac{\sec A \sec B}{\tan A - \tan B} = \csc (A - B)$

12. $\tan 2A = \dfrac{2}{\cot A - \tan A}$

13. $\dfrac{1 + \tan^2 x}{1 - \cos^2 x} = \sec^2 x \csc^2 x$

14. $\dfrac{2}{\sec x} = \dfrac{2 - 2 \sin^2 x}{\cos x}$

15. $\dfrac{\sin A}{1 - \cos A} + \dfrac{\sin A}{1 + \cos A} = 2 \csc A$

16. $\dfrac{\sin (A + B) + \sin (A - B)}{\sin (A + B) - \sin (A - B)} = \dfrac{\tan A}{\tan B}$

17. $\sec^2 x + \csc^2 x = (\tan x + \cot x)^2$

REGENTS TUNE-UP: CHAPTER 8

Each of the questions in this section has appeared on a previous Course III Regents Examination. Here is an opportunity for you to review Chapter 8 and, at the same time, prepare for the Course III Regents Examination.

1. Find a value of θ in the interval $0° \leq \theta < 360°$ that satisfies the equation $\sin^2 \theta - \sin \theta - 2 = 0$.

2. For all values of A for which the expression is defined, $\dfrac{\cot A}{\csc A}$ is equivalent to
   (1) $\cos A$　　(2) $\sin A$　　(3) $\dfrac{1}{\cos A}$　　(4) $\dfrac{1}{\sin A}$

3. If $\cos A = \dfrac{4}{5}$, find the positive value of $\tan \dfrac{A}{2}$.

4. What is the value of $\sin 210° \cos 30° - \cos 210° \sin 30°$?

5. If $\tan A = \dfrac{2}{3}$ and $\tan B = 3$, express $\tan (A - B)$ as a fraction in simplest form.

6. If $\sin x = \dfrac{5}{6}$, what is the value of $\cos 2x$?

7. If $\cos A = \dfrac{1}{2}$ and $\angle A$ is a positive acute angle, find the value of $\sin \dfrac{1}{2}A$.

8. The expression $\sin 2\theta \csc \theta$ is equivalent to
   (1) $\tan \theta$　　(2) $\tan 2\theta$　　(3) $2 \cos \theta$　　(4) $2 \sin \theta$

9. If $\sin A = \dfrac{3}{5}$ and $\sin B = \dfrac{2}{3}$, and $\angle A$ and $\angle B$ are acute angles, what is the value of $\cos (A - B)$?
   (1) $-\dfrac{2}{3}$　　(2) $\dfrac{4\sqrt{5}-6}{15}$　　(3) $\dfrac{4\sqrt{5}+2}{5}$　　(4) $\dfrac{4\sqrt{5}+6}{15}$

10. The expression $\cos^2 40 - \sin^2 40$ has the same value as
    (1) $\sin 20$　　(2) $\sin 80$　　(3) $\cos 80$　　(4) $\cos 20$

11. The expression $\dfrac{\sin x \cdot \cos x}{\tan x}$ is equivalent to

(1) 1   (2) $\sin^2 x$ (3) $\cos x$ (4) $\cos^2 x$

12. If x is a positive acute angle and $\cos x = \dfrac{1}{9}$, what is the value of $\cos\dfrac{1}{2}x$?

(1) $\dfrac{2}{3}$ (2) $\dfrac{1}{3}$ (3) $\dfrac{2\sqrt{5}}{3}$ (4) $\dfrac{\sqrt{5}}{3}$

13. Circle O has its center at the origin, OB = 1, and $\overline{BA} \perp \overline{OA}$. If m$\angle BOA = \theta$, which line segment shown has a length equal to cos θ?

14. If $\tan x = -\dfrac{24}{7}$ and x is an angle in

Quadrant II, find $\sin \dfrac{1}{2}x$.

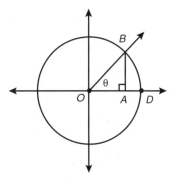

15. Solve for the *smallest* nonnegative value

of θ:

Exercise 13

$$\sqrt{3}\cos\theta + 1 = 2$$

16. If θ is in Quadrant II and $\cos \theta = -\dfrac{3}{4}$, find an exact value for $\sin 2\theta$.

17. **(a)** Find all values of x in the interval $0° \le x < 360°$ that satisfy the equation $\cos x \tan x + \cos x = 0$.

    (b) Express in simplest form:

$$\frac{64 - \cos^2 x}{\cos^2 x + 8\cos x} \div \frac{2\cos x - 16}{8\cos x}$$

18. **(a)** Find all values of θ in the interval $0° \le \theta < 360°$ that satisfy the equation $\cos 2\theta - \cos \theta = 0$.

    (b) For all values of 0 for which the expressions are defined, prove that the following is an identity:

$$\frac{2\sin^2\theta}{\sin 2\theta} + \frac{1}{\tan\theta} = \sec\theta \csc\theta.$$

19. Find, to the *nearest degree,* all values of θ in the interval $0° \le \theta < 360°$ that satisfy the equation $7\cos\theta + 1 = 6\sec\theta$.

20. Find, to the *nearest degree,* all values of θ in the interval $0° \le \theta < 360°$ that satisfy the equation $\sec^2\theta = 4 - 2\tan\theta$.

21. (a) For all values of θ for which the expressions are defined, prove that the following is an identity:

$$\frac{\cos 2\theta}{\sin \theta} + \sin \theta = \frac{\cot \theta}{\sec \theta}.$$

(b) Solve for $\cos \theta$: $\cos \left(\dfrac{\theta}{2}\right) = \dfrac{\sqrt{5}}{4}$.

22. For all values of θ for which the expressions are defined, prove that the following is an identity:

$$\frac{\sin 2\theta + \sin \theta}{\cos 2\theta + \cos \theta + 1} = \tan \theta.$$

23. Find, correct to the *nearest 10 minutes,* all values of x between $0°$ and $360°$ that satisfy the equation $2 \sin x + 4 \cos 2x = 3$.

ANSWERS TO SELECTED EXERCISES: CHAPTER 8

Section 8.1

1. $\dfrac{4}{5}$

2. $-\dfrac{13}{5}$

3. (3)

4. (4)

5. (1)

6. $\cos^2 A$

7. $\sec A$

8. $\tan^2 A$

9. $\dfrac{\cos^2 x}{\sin x}$

10. $\cos^2 A$

11. $\sin^2 A$

13. $\dfrac{\sin^2 x}{\cos^2 x}$

15. $\dfrac{1}{\cos x}$

17. $2 \sin x \cos x$

18. (a) $2a$ **(c)** $-b$
   (b) $2(a - b)$ **(d)** $-2a$

19. (1)

21. $\dfrac{\sin \theta \tan \theta + \cos \theta}{\cos \theta} = \dfrac{\sin \theta \tan \theta}{\cos \theta} + \dfrac{\cos \theta}{\cos \theta}$
$= \tan^2 \theta + 1 = \sec^2 \theta$

23. Since $\cos^2 x = 1 - \sin^2 x$ and $\angle x$ is acute, $\cos x = \sqrt{1 - \sin^2 x}$. Hence

$$\tan x = \frac{\sin x}{\cos x} = \frac{\sin x}{\sqrt{1 - \sin^2 x}}.$$

Section 8.2

1. 135°, 315°	**13.** 210°, 330°	**25.** 26.5°, 116.5°,
3. 210°, 330°	**15.** 60°, 180°, 300°	206.5°, 296.5°
5. 135°, 315°	**17.** 210°, 270°, 330°	**27.** 195.5°, 344.5°
7. 60°, 120°, 240°,	**19.** 124.8°, 304.8°	**29.** 70.7°, 160.7°,
300°	**21.** 19.5°, 160.5°	250.7°, 340.7°
9. 30°, 150°, 270°	**23.** 90°, 221.8°, 318.8°	**30.** 7.9°, 172.1°,
11. 90°, 270°		233.6°, 306.4°

Section 8.3

1. $\dfrac{1}{2}$ **2.** $\dfrac{\sqrt{2}}{2}$ **3.** 1   **4.** $\dfrac{1}{2}$ **5.** $-\dfrac{1}{8}$

7. **(a)** $-\dfrac{56}{65}$ **(b)** $-\dfrac{63}{65}$ **(c)** 7   **(d)** $-\dfrac{65}{33}$

9. $\tan 15° = \tan(45° - 30°) = \dfrac{\tan 45° - \tan 30°}{1 + (\tan 45°)(\tan 30°)} = \dfrac{1 - \dfrac{1}{\sqrt{3}}}{1 + \dfrac{1}{\sqrt{3}}} = \dfrac{\sqrt{3} - 1}{\sqrt{3} + 1}$

$= \dfrac{\sqrt{3} - 1}{\sqrt{3} + 1} \cdot \dfrac{\sqrt{3} - 1}{\sqrt{3} - 1} = \dfrac{4 - 2\sqrt{3}}{2} = 2 - \sqrt{3}$

11. $\dfrac{1 - a}{1 + a}$ **13.** (4)   **14.** (1)   **15.** (3)   **16.** (4)

Section 8.4

1. **(a)** $\dfrac{240}{289}$ **(b)** $\dfrac{161}{289}$ **(c)** $\dfrac{240}{161}$

3. **(a)** $-\dfrac{12}{13}$ **(b)** $\dfrac{5}{13}$ **(c)** $-\dfrac{12}{5}$

5. **(a)** 30°, 150°, 270°   **(d)** 22.5°, 112.5°, 202.5°, 292.5°
   (b) 0°, 120°, 180°, 240°   **(e)** 120°, 240°
   (c) 90°, 210°, 270°, 330°   **(f)** 15°

6. **(a)** 38°40′, 141°20′, 270°   **(b)** 104°30′, 255°30′

7. $\dfrac{\sin 2A}{2 \tan A} = \sin 2A \div 2 \tan A = 2 \sin A \cos A \times \dfrac{\cos A}{2 \sin A} = \cos^2 A$

9. $\sin 2x \cot 2x + 1 = (\sin 2x)\left(\dfrac{\cos 2x}{\sin 2x}\right) + 1 = \cos 2x + 1$

$$= (2\cos^2 x - 1) + 1 = 2\cos^2 x$$

11. $\dfrac{1 + \cos A + \cos 2A}{\sin A + \sin 2A} = \dfrac{1 + \cos A + 2\cos^2 A - 1}{\sin A + 2\sin A \cos A}$

$$= \dfrac{\cos A(1 + 2\cos A)}{\sin A(1 + 2\cos A)} = \cot A$$

Section 8.5

1. (a) 0.6 **2. (a)** $\dfrac{1}{3}$ **3. (a)** $\dfrac{2}{\sqrt{7}}$ **5. (a)** $-\dfrac{1}{3}$

(b) -0.8 **(b)** $\dfrac{\sqrt{8}}{3}$ **(b)** $\dfrac{2}{\sqrt{5}}$ **(b)** $\dfrac{1}{\sqrt{2}}$

(c) -0.75 **(c)** $\dfrac{1}{\sqrt{8}}$ **4. (a)** 0.9 **(b)** $\dfrac{7}{\sqrt{58}}$ **6.** -0.68

Section 8.6

1. $\sin^2 A(1 + \tan^2 A) = \sin^2 A(\sec^2 A) = \dfrac{\sin^2 A}{\cos^2 A} = \tan^2 A$

3. $\dfrac{2\cot A}{1 + \cot^2 A} = \dfrac{2\cot A}{\csc^2 A} = 2\cot A \sin^2 A = 2\dfrac{\cos A}{\sin A}\sin^2 A$

$$= 2\sin A \cos A = \sin 2A$$

5. $\dfrac{\sin 2A}{1 + \cos 2A} = \dfrac{2\sin A \cos A}{1 + (2\cos^2 A - 1)} = \dfrac{2\sin A \cos A}{2\cos^2 A} = \dfrac{\sin A}{\cos A} = \tan A$

7. $\dfrac{2\tan\theta}{1 + \tan^2\theta} = \dfrac{2\tan\theta}{\sec^2\theta} = 2\tan\theta\cos^2\theta = 2\dfrac{\sin\theta}{\cos\theta}\cos^2\theta = 2\sin\theta\cos\theta = \sin 2\theta$

9. Since $\dfrac{(\cos x + \sin x)^2}{1 + \sin 2x} = \dfrac{\cos^2 x + \sin^2 x + 2\sin x \cos x}{1 + \sin 2x} = \dfrac{1 + \sin 2x}{1 + \sin 2x} = 1$

and $\dfrac{\cos x \tan x}{\csc x} = (\cos x)\left(\dfrac{\sin x}{\cos x}\right)(\sin x) = 1,$

then $\dfrac{(\cos x + \sin x)^2}{1 + \sin 2x} = \dfrac{\cos x \tan x}{\csc x}.$

11.

$\dfrac{\sec A \sec B}{\tan A - \tan B} = \dfrac{\dfrac{1}{\cos A \cos B}}{\dfrac{\sin A}{\cos A} - \dfrac{\sin B}{\cos B}} = \dfrac{1}{\sin A \cos B - \sin B \cos A}$

$$= \dfrac{1}{\sin(A - B)} = \csc(A - B)$$

13. $\dfrac{1+\tan^2 x}{1-\cos^2 x} = \dfrac{\sec^2 x}{\sin^2 x} = \sec^2 x \csc^2 x$

15. $\dfrac{\sin A}{1-\cos A} + \dfrac{\sin A}{1+\cos A} = \sin A\left[\dfrac{(1+\cos A)+(1-\cos A)}{(1-\cos A)(1+\cos A)}\right]$

$$= \dfrac{2\sin A}{1-\cos^2 A} = \dfrac{2\sin A}{\sin^2 A} = \dfrac{2}{\sin A} = 2\csc A$$

17. $(\tan x + \cot x)^2 = \tan^2 x + 2\tan x \cot + \cot^2 x$
$$= \tan^2 x + 2 + \cot^2 x$$
$$= (\tan^2 x + 1) + (1 + \cot^2 x)$$
$$= \sec^2 \quad x \quad + \quad \csc^2 x$$

Regents Tune-Up: Chapter 8

1. 270°

2. (1)

3. $\dfrac{1}{3}$

4. 0

5. $-\dfrac{7}{9}$

6. $-\dfrac{14}{36}$

7. $\dfrac{1}{2}$

8. (3)

9. (4)

10. (3)

11. (4)

12. (4)

13. $\overline{OA}$

14. $\dfrac{4}{5}$

15. 0°

16. $-\dfrac{3\sqrt{7}}{8}$

17. (a) 90°, 135°, 270°, 315°

(b) −4

18. (a) 0°, 120°, 240°

(b) $\dfrac{2\sin^2\theta}{\sin 2\theta} + \dfrac{1}{\tan\theta} = \dfrac{2\sin^2\theta}{2\sin\theta\cos\theta} + \dfrac{\cos\theta}{\sin\theta} = \dfrac{\sin\theta}{\cos\theta} + \dfrac{\cos\theta}{\sin\theta}$

$$= \dfrac{\sin^2\theta + \cos^2\theta}{\cos\theta\sin\theta} = \dfrac{1}{\cos\theta}\cdot\dfrac{1}{\sin\theta} = \sec\theta\csc\theta$$

19. 31°, 180°, 329°

20. 45°, 108°, 225°, 288°

21. (a) $\dfrac{\cos 2\theta}{\sin\theta} + \sin\theta = \dfrac{\cos 2\theta + \sin^2\theta}{\sin\theta} = \dfrac{1 - 2\sin^2\theta + \sin^2\theta}{\sin\theta}$

$$= \dfrac{1 - \sin^2\theta}{\sin\theta} = \dfrac{\cos^2\theta}{\sin\theta} = \dfrac{\cos\theta}{\sin\theta}\cos\theta$$

$$= \cot\theta\dfrac{1}{\sec\theta} = \dfrac{\cot\theta}{\sec\theta}$$

(b) $-\dfrac{3}{8}$

22. $\dfrac{\sin 2\theta + \sin\theta}{\cos 2\theta + \cos\theta + 1} = \dfrac{2\sin\theta\cos\theta + \sin\theta}{2\cos^2\theta - 1 + \cos\theta + 1} = \dfrac{2\sin\theta(\cos\theta + 1)}{2\cos\theta(\cos\theta + 1)} = \tan\theta$

23. 30°, 150°, 194°30′, 345°30′

CHAPTER 9
SOLVING TRIANGLES

9.1 FINDING THE AREA OF A TRIANGLE, GIVEN SAS

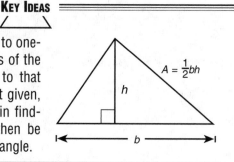

KEY IDEAS

The **area** of a triangle is equal to one-half the product of the lengths of the base and the altitude drawn to that base. When the altitude is not given, trigonometry may be helpful in finding the altitude, which can then be used to find the area of the triangle.

$A = \frac{1}{2}bh$

Area of a Triangle, Given SAS

In Figure 9.1, the area of $\triangle ABC$ may be expressed as $\frac{1}{2}bh$. In right triangle BDC, $\sin C = \frac{h}{a}$ or, equivalently, $h = a \sin C$. This means that h can be replaced by $a \sin C$ in the area formula:

$$\text{Area } \triangle ABC = \frac{1}{2}bh$$

$$= \frac{1}{2}b(a \sin C)$$

$$= \frac{1}{2}ab \sin C$$

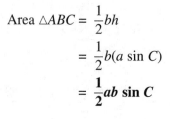
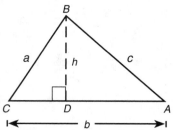

Figure 9.1   Finding the Area of a Triangle, Given SAS

AREA OF A TRIANGLE

The area of a triangle equals one-half the product of the lengths of any two sides and the sine of the included angle. If K represents the area of $\triangle ABC$, then

$$K = \frac{1}{2}ab \sin C = \frac{1}{2}ac \sin B = \frac{1}{2}bc \sin A$$

Examples

1. If $m\angle A = 150$, $b = 8$, and $c = 10$, find the area of $\triangle ABC$.

Solution: In $\triangle ABC$, b and c represent the lengths of a pair of adjacent sides and $\angle A$ is the included angle.

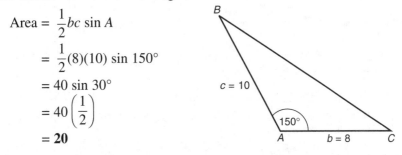

$$\text{Area} = \frac{1}{2}bc \sin A$$

$$= \frac{1}{2}(8)(10) \sin 150°$$

$$= 40 \sin 30°$$

$$= 40\left(\frac{1}{2}\right)$$

$$= \mathbf{20}$$

2. In parallelogram $ABCD$, $AB = 20$, $AD = 10$, and $m\angle A = 45$. Find the *exact* area of parallelogram $ABCD$.

Solution: The area of a parallelogram is equal to the product of the base and the altitude drawn to that base. In parallelogram $ABCD$, draw $\overline{DE}$ perpendicular to $\overline{AB}$, as shown in the accompanying diagram. In right triangle AED,

$$\sin 45° = \frac{DE}{10} \quad \text{or} \quad DE = 10 \sin 45°$$

$$= 10 \frac{\sqrt{2}}{2}$$

$$= 5\sqrt{2}$$

$$\begin{aligned}
\text{Area} &= \text{base} \times \text{altitude}\\
&= AB \times DE\\
&= 20 \times 5\sqrt{2}\\
&= \mathbf{100\sqrt{2}}
\end{aligned}$$

Note that:
- Since the *exact* area is required, the answer *must* be expressed in radical form.
- The problem could also have been solved by finding the sum of the areas of the two congruent triangles formed by drawing diagonal $\overline{DB}$.

3. Find the value of sin C if the area of $\triangle ABC$ is 15, $a = 5$, and $b = 10$.

Solution:

$$\text{Area} = \frac{1}{2}ab \sin C$$

$$15 = \frac{1}{2}(5)(10) \sin C$$

Multiply each side by 2: $30 = 50 \sin C$

Divide by 50: $\dfrac{30}{50} = \sin C$

$$\sin C = \frac{3}{5} \text{ or } \mathbf{0.6}$$

Exercise Set 9.1

1. Find the area of $\triangle ABC$ if $b = 7$, $c = 12$, and $\angle A$ measures:
(a) 150°   (b) 45°   (c) 60°   (d) 90°

2. Find, to the *neareast square inch,* the area of a triangle if the lengths of two sides are 10 inches and 20 inches and the measure of their included angle is:
(a) 28°   (b) 65°   (c) 100°   (d) 118°40′

3. Find the area of $\triangle DEF$ if $d = 4$, $e = 7$, and $\sin F = 0.3$.

4. Find the value of sin C if the area of $\triangle ABC$ is 9, $a = 6$, and $b = 20$.

5. In parallelogram $ABCD$, $AB = 12$, $AD = 5$, and $m\angle A = 60$. Find the exact area of the parallelogram.

6. The length of each of the legs of an isosceles triangle is 6 inches, and the measure of each base angle is 75°. Find the area of the triangle, correct to the *nearest integer.*

7. The area of an isosceles triangle is 16 square units. If the measure of the vertex angle is 150, find the length of each leg of the triangle.

8. The area of an equilateral triangle is $25\sqrt{3}$. What is the length of a side of the triangle?

9. Georgina wants to build a garden in the shape of an isosceles triangle with the length of one of the congruent sides equal to 12 yards. If the area of the garden will be 55 square yards, find, to the *nearest tenth of a degree*, the greatest possible measure of the vertex angle of the triangle.

9.2 SOLVING TRIANGLES: THE LAW OF SINES

KEY IDEAS

Solving a triangle means finding the measures of its unknown angles or sides by using other parts of the triangle whose measures are given.

The **Law of Sines** establishes a proportion involving the lengths of any two sides of a triangle and the *sines* of the angles opposite these sides. Thus, the Law of Sines is used when a relationship involving the measures of two sides and two angles of a triangle is needed.

Law of Sines

In any triangle the ratio of the length of each side to the sine of the angle opposite that side is the same.

MATH FACTS

LAW OF SINES

In $\triangle ABC$, each of the following proportions is true:

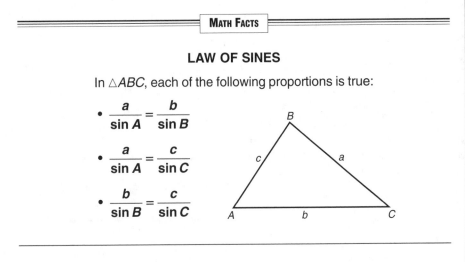

- $\dfrac{a}{\sin A} = \dfrac{b}{\sin B}$

- $\dfrac{a}{\sin A} = \dfrac{c}{\sin C}$

- $\dfrac{b}{\sin B} = \dfrac{c}{\sin C}$

Examples

1. In $\triangle ABC$, $a = 12$, $\sin A = 0.6$, and $\sin B = 0.4$. Find b.

Solution: Since SAA (**S**ide-**A**ngle-**A**ngle) is given, solve for b using the Law of Sines:

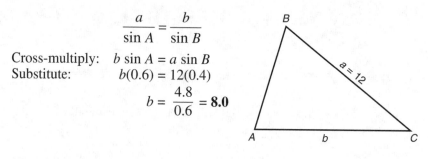

$$\frac{a}{\sin A} = \frac{b}{\sin B}$$

Cross-multiply: $b \sin A = a \sin B$

Substitute: $b(0.6) = 12(0.4)$

$$b = \frac{4.8}{0.6} = \textbf{8.0}$$

2. In $\triangle RST$, $r = 6$, $t = 9$, and $m\angle R = 30$. Find the value of $\sin T$.

Solution: Use Law of Sines:

$$\frac{r}{\sin R} = \frac{t}{\sin T}$$

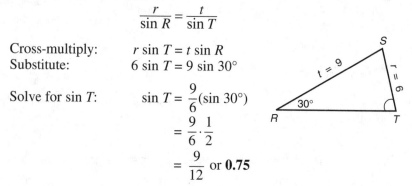

Cross-multiply: $r \sin T = t \sin R$

Substitute: $6 \sin T = 9 \sin 30°$

Solve for $\sin T$:

$$\sin T = \frac{9}{6}(\sin 30°)$$

$$= \frac{9}{6} \cdot \frac{1}{2}$$

$$= \frac{9}{12} \text{ or } \textbf{0.75}$$

3. In $\triangle ABC$, $m\angle A = 59$, $m\angle B = 74$, and $b = 100$. Find c to the *nearest tenth*.

Solution: Solving for c using the Law of Sines requires knowledge of $m\angle C$. Since the sum of the degree measures of the angles of a triangle is 180,

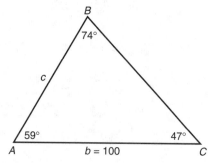

$$m\angle C = 180 - (m\angle A + m\angle B)$$
$$= 180 - (59 \quad + 74)$$
$$= 47$$

Now apply the Law of Sines:

$$\frac{b}{\sin B} = \frac{c}{\sin C}$$
$$c \sin B = b \sin C$$
$$c \sin 74° = 100 \sin 47°$$
$$c \approx \frac{100(0.7314)}{0.9613} \approx 76.08$$

Thus, to the *nearest tenth*, $c = \textbf{76.1}$.

4. In $\triangle ABC$, $\angle A$ measures $26°10'$, $a = 88.2$, $b = 100$, and $\angle B$ is acute. Find the measure of $\angle C$ correct to the *nearest 10 minutes* or the *nearest tenth of a degree.*

Solution: First solve for $\angle B$ using the Law of Sines.

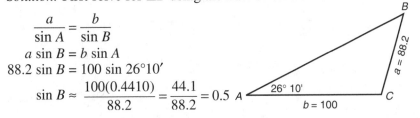

$$\frac{a}{\sin A} = \frac{b}{\sin B}$$
$$a \sin B = b \sin A$$
$$88.2 \sin B = 100 \sin 26°10'$$
$$\sin B \approx \frac{100(0.4410)}{88.2} = \frac{44.1}{88.2} = 0.5$$

If $\sin B = 0.5$, then $\angle B$ is $30°$ or $150°$. It is given that $\angle B$ is acute, so $\angle B$ is $30°$. Since the sum of the measures of the angles of a triangle is $180°$,

$$\angle C = 180° - (26°10' + 30°)$$
Convert $1°$ to $60'$: $\qquad = 179°60' - 56°10'$
$$= \mathbf{123°50'} \text{ or } \mathbf{123.8°}$$

Solving "Double-Triangle" Problems

Sometimes two triangles overlap and have angles or sides in common. In order to apply a trigonometric relationship to one triangle, it may be necessary first to solve for a shared angle or common side by working in the other triangle.

Example

5. The angle of elevation from a ship at point A to the top of a lighthouse, point B, is $43°$. When the ship reaches point C, 300 meters closer to the lighthouse, the angle of elevation is $56°$. Find, to the *nearest meter,* the height, BD, of the lighthouse.

Solution: Reason "backwards": In right angle BDC, if the length of hypotenuse $\overline{BC}$ were known, BD could be determined by using the sine ratio. To solve for BC, use the Law of Sines in $\triangle ABC$. This requires finding the measure of another angle of this triangle.

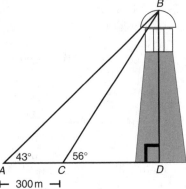

- Find m∠ABC. Since the measure of exterior angle DCB of △ABC is equal to the sum of the measures of the two nonadjacent angles of this triangle, m∠ABC + 43 = 56. This means that m∠ABC = 13.
- In △ABC, find BC using the Law of Sines.

$$\frac{300}{\sin 13°} = \frac{BC}{\sin B°} \quad \text{or} \quad BC \sin 13° = 300 \sin 43°$$

$$BC = \frac{300 \sin 43°}{\sin 13°}$$

$$\approx \frac{300(0.6820)}{0.2250} \approx 909.33$$

- In △BDC, find BD using the sine ratio.

$$\sin 56° \approx \frac{BD}{909.33} \quad \text{or} \quad BD \approx 909.33(\sin 56°)$$

$$\approx 753.86874$$

$$\approx 754 \text{ to the } nearest\ meter$$

Exercise Set 9.2

1. In △ABC, $a = 5$, $b = 9$, and $\sin A = 0.30$. Find the value of $\sin B$.

2. In △ABC, $\sin A = \frac{2}{3}$, $\sin B = \frac{2}{5}$, and $a = 15$. Find b.

3. In △RST, $\sin R = 0.36$, $r = 12$, and $s = 18$. Find the value of $\sin S$.

4. In △ABC, $A = 30°$, $C = 105°$, and $b = 6$. Find a.

5. In △DEF, $\sin D = 0.15$ and $\sin E = 0.60$. What is the ratio of side d to side e?

6. An ice skater starts from point A and skates in a straight line across a frozen lake to point B, which is 630 meters away. Then she skates directly to point C. If the measure of ∠ABC is 47°30′ and the measure of ∠CAB is 31°20′, find, to the *nearest meter,* the distance from C to A.

7. The measures of the angles of a triangle are in the ratio 2 : 3 : 4. The shortest side of the triangle is 30 feet. Find, to the *nearest foot,* the length of the *longest* side.

8. The measures of two angles of a triangle are 46°40′ and 58°30′, and the length of the included side is 64.3 feet. Find, to the *nearest tenth of a foot,* the length of the *shortest* side of the triangle.

9. In $\triangle ABC$, the measure of $\angle A$ is $42°20'$, $a = 18$, $b = 25$, and $\angle B$ is acute. Find the measure of $\angle C$ to the *nearest 10 minutes*.

10. Just as a plane flies over a level fence joining two ground observation posts 4680 feet apart, it is spotted by observers at both posts. If the angles of elevation of the plane from the two posts at this moment are $72°20'$ and $51°50'$, respectively, find, to the *nearest 10 feet,* the height at which the plane is flying.

11. In the accompanying diagram of $\triangle ABC$, $m\angle A = 65$, $m\angle B = 70$, and the length of the side opposite vertex B is 7 inches. Find, to the *nearest hundredth*, the number of square inches in the area of $\triangle ABC$.

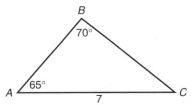

9.3 SOLVING TRIANGLES: THE LAW OF COSINES

KEY IDEAS

The **Law of Cosines** gives a relationship between the three sides of a triangle and the cosine of one of its angles. The Law of Cosines can be used to solve a triangle in which (1) the measures of any two sides and the included angle (*SAS*) are known or (2) the lengths of the three sides (*SSS*) are known.

Using the Law of Cosines, Given SAS

The Law of Cosines states that in *any* triangle the square of the length of one side is equal to the sum of the squares of the lengths of the other two sides minus twice the product of these two side lengths and the cosine of the included angle. The chart below gives three equivalent forms of the Law of Cosines.

Triangle	Given	To Find	Use Law of Cosines
	b, $\angle A$, c	a	$a^2 = b^2 + c^2 - 2bc \cos A$
	a, $\angle B$, c	b	$b^2 = a^2 + c^2 - 2ac \cos B$
	a, $\angle C$, b	c	$c^2 = a^2 + b^2 - 2ab \cos C$

The Law of Cosines formula simplifies to the Pythagorean theorem when it is used in a right triangle to find the length of the hypotenuse.

In Figure 9.2,

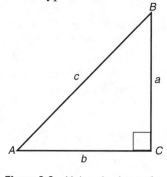

$$c^2 = a^2 + b^2 - 2ab \cos C$$
$$= a^2 + b^2 - 2ab \cos 90°$$
$$= a^2 + b^2 - 2ab(0)$$
$$= a^2 + b^2 - 0$$
$$= a^2 + b^2$$

Thus the Law of Cosines may be viewed as a generalization of the Pythagorean theorem. The Law of Cosines, unlike the Pythagorean theorem, can be applied to *any* triangle to find the length of a side, given the measures of the other two sides and the included angle (SAS).

Figure 9.2 Using the Law of Cosines in a Right Triangle

Examples

1. In $\triangle ABC$, $a = 4$, $b = 5$, and $\cos C = \dfrac{1}{8}$. Find the length of side c.

Solution: $c^2 = a^2 + b^2 - 2ab \cos C$

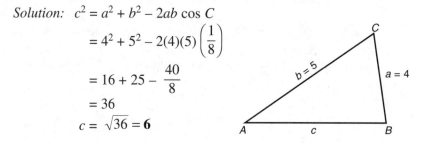

$$= 4^2 + 5^2 - 2(4)(5)\left(\dfrac{1}{8}\right)$$
$$= 16 + 25 - \dfrac{40}{8}$$
$$= 36$$
$$c = \sqrt{36} = 6$$

2. In $\triangle RST$, $s = 3$, $t = 5$, and $m\angle R = 120$. Find the length of side r.

Solution: Since sides s and t and their included angle, R, are given, solve for r using the Law of Cosines:

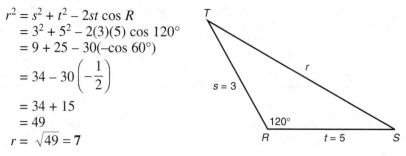

$$r^2 = s^2 + t^2 - 2st \cos R$$
$$= 3^2 + 5^2 - 2(3)(5) \cos 120°$$
$$= 9 + 25 - 30(-\cos 60°)$$
$$= 34 - 30\left(-\dfrac{1}{2}\right)$$
$$= 34 + 15$$
$$= 49$$
$$r = \sqrt{49} = 7$$

3. The lengths of two adjacent sides of parallelogram *ABCD* are 16 centimeters and 20 centimeters. The measure of the angle between the two sides is 35°. Find, to the *nearest centimeter,* the length of the *longer* diagonal.

Solution: In parallelogram *ABCD,* let *AB* = 16, *AD* = 20, and m∠*A* = 35. The longer diagonal of a parallelogram lies opposite the largest angle of the parallelogram. Since consecutive angles of a parallelogram are supplementary, m∠*B* = 180 − 35 = 145. Hence $\overline{AC}$ is the longer diagonal of the parallelogram.

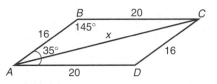

Let *x* = *AC*. Since opposite sides of a parallelogram have the same length, *BC* = 20. Find *x* by applying the Law of Cosines in △*ABC*:

$$x^2 = (AB)^2 + (BC)^2 - 2(AB)(BC) \cos 145°$$
$$= 16^2 + 20^2 - 2(16)(20)(-\cos 35°)$$
$$\approx 256 + 400 + 640(0.8192)$$
$$\approx 1180.3$$
$$x \approx \sqrt{1180.3} \approx \textbf{34} \text{ to the } \textit{nearest centimeter}$$

Using the Law of Cosines, Given SSS

If the lengths of the three sides of △*ABC* are known (SSS), then any angle of the triangle can be solved for by using one of the following equivalent forms of the Law of Cosines:

$\cos A = \dfrac{b^2 + c^2 - a^2}{2bc}$	$\cos B = \dfrac{a^2 + c^2 - b^2}{2ac}$	$\cos C = \dfrac{a^2 + b^2 - c^2}{2ab}$

Example

4. In △*ABC*, *a* = 4, *b* = 8, and *c* = 5. Find, correct to the *nearest degree,* the measure of the *largest* angle of the triangle.

Solution: In a triangle, the angle having the greatest measure lies opposite the side having the greatest length. Since ∠*B* is the largest angle of △*ABC*, choose the form of the Law of Cosines that solves for cos *B*. Let *a* = 4, *b* = 8, and *c* = 5.

$$\cos B = \frac{a^2 + c^2 - b^2}{2ac}$$

$$= \frac{4^2 + 5^2 - 8^2}{2(4)(5)}$$

$$= \frac{16 + 25 - 64}{40}$$

$$= -\frac{23}{40} = -0.5750$$

Since $\cos B$ is negative, $\angle B$ is obtuse. The reference angle is the angle whose cosine is closest to 0.5750. Using a scientific calculator, find that this angle is 55°. Thus

$$\angle B = \cos^{-1}(-0.5750) = 180° - 55° = \mathbf{125°}.$$

Exercise Set 9.3

1. In $\triangle ABC$, $b = 8$, $c = 6$, and $\cos A = \dfrac{17}{32}$. Find a.

2. In $\triangle ABC$, $a = 4$, $b = 3$, and $\cos C = -\dfrac{1}{2}$. What is the length of c?

   (1) 7   (2) $\sqrt{13}$ (3) $\sqrt{37}$ (4) $\sqrt{19}$

3. In $\triangle DEF$, $d = 5$, $e = 7$, and $f = 8$. Find the value of the cosine of the *largest* angle of the triangle.

4. In $\triangle ABC$, $a = 6$, $b = 7$, and $c = 8$. What is $\cos A$ in simplest fractional form?

   (1) $\dfrac{3}{16}$ (2) $\dfrac{11}{16}$ (3) $\dfrac{77}{96}$ (4) $\dfrac{51}{112}$

5. The lengths of the sides of a triangle are 6, 10, and 14. Find the value of the cosine of the *smallest* angle of the triangle.

6. In $\triangle ABC$, $a = \sqrt{19}$, $b = 3$, and $c = 2$. Find the measure of $\angle A$.

7. In $\triangle ABC$, $a = 5$ and $b = 6$. Express $\cos C$ in terms of side c.

8. From point P, tangents $\overline{PA}$ and $\overline{PB}$ are drawn to circle O in such a way that $\angle APB$ measures 120°. If the length of chord $\overline{AB}$ is 9, find the length of $\overline{PA}$.

9. The sides of a triangle have lengths 58, 92, and 124.
(a) Find, to the *nearest ten minutes,* the *largest* angle of the triangle.
(b) Find, to the *nearest integer,* the area of the triangle.

10. Points *A* and *B* are separated by an obstacle. To find the distance between them, a third point, *C*, is selected, which is 50 meters from *A* and 75 meters from *B*. If the measure of ∠*BAC* is 120°40′, find, to the *nearest meter,* the distance from *A* to *B*.

11. The lengths of two consecutive sides of a parallelogram are 8 centimeters and 10 centimeters. If the length of the longer diagonal of the parallelogram is 14 centimeters, find, to the *nearest 10 minutes:*
(a) the *largest* angle of the parallelogram
(b) the angle formed by the longer side and the diagonal

12. Main Street and Park Avenue intersect at an angle of 74°. Mr. Jones lives on Main Street, 50 meters from the intersection, and Mr. Diaz lives on Park Avenue, 40 meters from the intersection. The triangle formed by the intersection and the houses is an acute triangle. Find, to the *nearest tenth of a meter,* the distance between the two houses.

13. A metal frame is constructed in the form of an isosceles trapezoid. Each base angle measures 73°30′, the length of the shorter base is 8.0 feet, and the length of each of the nonparallel sides is 5.0 feet. Find, to the *nearest tenth of a foot,* the length of a diagonal brace of this frame.

14. The lengths of two adjacent sides of a parallelogram are 50 and 80 inches. The length of the longer diagonal of the parallelogram is 100 inches.
(a) Find, to the *nearest tenth of a degree,* the measure of the *largest* angle of the parallelogram.
(b) Using the answer obtained in part (a), find, to the *nearest square inch,* the area of the parallelogram.

15. In the accompanying diagram, m∠*C* = 90, m∠*CBD* = 60, *AB* = 10, *BC* = 10, and *AD* = 15.
(a) Find m∠*A* to the *nearest degree.*
(b) Using the answer obtained in part (a), find the area of quadrilateral *ABCD* to the *nearest square unit.*

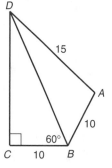

9.4 SOLVING TRIANGLES: MORE APPLICATIONS

∧
KEY IDEAS
∠‾‾∖

In some situations, the sides of a triangle are also parts of a parallel-ogram. In other problems, the sides of a triangle are chords, secants, or tangents of a circle. In order to apply the Law of Sines or the Law of Cosines to these triangles, it may be necessary first to determine the measure of a part of the triangle, using geometric relationships found in parallelograms or circles.

Parallelogram of Forces

A single force that has the same effect as two forces acting simultane-ously is called a **resultant force.** When the two forces form an acute or an obtuse angle, the resultant force can be represented by the diagonal of a parallelogram whose adjacent sides represent the two forces. Figure 9.3 shows two forces being applied at point A at an angle of 60°. If $\overline{AB}$

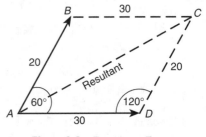

Figure 9.3 Resultant Force

is a 20-pound force and $\overline{AD}$ is a 30-pound force, then diagonal $\overline{AC}$ represents the resultant force.

Its magnitude can be determined by applying the Law of Cosines in $\triangle ADC$ (or $\triangle ABC$). Since opposite sides of a parallelogram are equal, $AB = DC = 20$. The measure of $\angle D$ is 120° since consecutive angles of a parallel-ogram are supplementary. Thus

$$(AC)^2 = (AD)^2 + (DC)^2 - 2(AD)(DC) \cos D$$
$$= (30)^2 + (20)^2 - 2(30)(20) \cos 120°$$
$$= 900 + 400 - 1200 \left(-\frac{1}{2}\right)$$
$$= 1900$$
$$AC = \sqrt{1900} \approx 43.6 \text{ to the } \textit{nearest tenth}$$

Hence the magnitude of the resultant force is **43.6 pounds.**

Example 1 illustrates that it may be necessary to use the Law of Sines in solving a "parallelogram of forces" problem.

Example

1. Two forces of 437 pounds and 876 pounds act upon a body at an acute angle with each other. The angle between the resultant force and the 437-pound force is $41°10'$. Find, to the *nearest 10 minutes,* the angle formed by the 437-pound and the 876-pound forces.

Solution: Draw parallelogram *ABCD*, where $AB = 437$, $AD = 876$, and the measure of $\angle CAB = 41°10'$. To find the measure of $\angle DAB$, let $x = $ m$\angle DAC$, so

$$\angle DAB = 41°10' + x.$$

In a parallelogram, opposite sides have the same measure, as do alternate interior angles. Hence

$$BC = AD = 876$$

and

$$m\angle BCA = m\angle DAC = x.$$

To find x, apply the Law of Sines in $\triangle ABC$. Then use your scientific calculator to help perform the calculations.

$$\sin x = \frac{437}{876} \times \sin 41°10'$$
$$\approx \frac{437}{876} \times \sin 41.167°$$
$$\approx 0.3284$$
$$x \approx \sin^{-1} 0.3284$$
$$\approx 19.17°$$
$$\approx 19°10'$$

Since $\angle DAB = 41°10' + 19°10' = 60°20'$, the angle formed by the two forces is **60°20′**, correct to the *nearest 10 minutes.*

Solving Triangles That Intersect Circles

Sometimes circle measurement relationships must be used before the Law of Sines or the Law of Cosines can be applied.

Example

2. In the accompanying diagram, $\overleftrightarrow{AB}$ is tangent to circle O at B, $\overleftrightarrow{AB}$ and $\overleftrightarrow{ADC}$ intersect at A, and chord $\overline{BC} \cong$ chord $\overline{CD}$. If m$\widehat{BD} = 60$ and $BC = 8$, find the length of tangent segment $\overline{AB}$.

Solution: $\overline{AB}$ is a side of $\triangle ABC$ that can be solved for using the Law of Sines if SAA (side-angle-angle) is known. It is given that side $BC = 8$. Angle C is an inscribed angle, so

$$m\angle C = \frac{1}{2}(60) = 30.$$

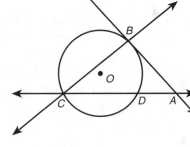

The measure of $\angle A$ can be determined by first finding the measure of arc BC.

In a circle, congruent chords intercept equal arcs, so let

$$x = m\widehat{BC} = m\widehat{CD}.$$

Since the sum of the degree measures of the arcs of a circle is 360,

$$x + x + 60 = 360$$
$$2x = 300$$
$$m\widehat{BC} = x = 150$$

Angle A is formed by a tangent and a secant, so it is measured by one-half the difference of the measures of its intercepted arcs. Since $m\widehat{BC} = 150$ and $m\widehat{BD} = 60$,

$$m\angle A = \frac{1}{2}(150 - 60) = 45.$$

Applying the Law of Sines in $\triangle ABC$ gives

$$\frac{BC}{\sin A} = \frac{AB}{\sin C} \quad \text{or} \quad AB \sin A = BC \sin C$$

$$AB(\sin 45°) = 8 \quad \sin 30°$$

$$AB\left(\frac{\sqrt{2}}{2}\right) = 8\left(\frac{1}{2}\right)$$

$$AB = \frac{2}{\sqrt{2}}(4) = \frac{8}{\sqrt{2}}$$

The length of tangent segment AB is $\dfrac{8}{\sqrt{2}}$ or, rationalizing the denominator,

$$AB = \frac{8}{\sqrt{2}} \cdot \frac{\sqrt{2}}{\sqrt{2}} = 4\sqrt{2}.$$

3. Using the diagram provided in Example 2, if $\widehat{BC} \cong \widehat{CD}$, $AB = 6$, and $AD = 4$, find:

(a) the length of: *(1)* $\overline{CD}$ *(2)* $\overline{BC}$ *(3)* $\overline{ADC}$

(b) $\angle A$ correct to the *nearest tenth of a degree*

Solutions: **(a)** If a tangent segment and a secant are drawn to a circle from the same point, then the square of the length of the tangent segment is equal to the product of the lengths of the secant segment and its external segment. Thus

$$(AD)(ADC) = (AB)^2$$
$$(4)(4 + CD) = (6)^2$$
$$4(4 + CD) = 36$$
$$4 + CD = 9$$
$$CD = 5$$

and

$$ADC = CD + AD = 5 + 4 = 9$$

If $CD = 5$, then BC also equals 5, since congruent arcs have equal chords. Thus

$$(1) \; CD = 5 \qquad (2) \; BC = 5 \qquad (3) \; ADC = 9$$

(b) In $\triangle ABC$, the lengths of the three sides are known, so the measure of $\angle A$ can be found by using the Law of Cosines.

$$(BC)^2 = (AB)^2 + (ADC)^2 - 2(AB)(ADC) \cos A$$
$$5^2 = 6^2 \quad + 9^2 \quad\quad - 2(6)(9) \cos A$$
$$25 = 36 \quad + 81 \quad\quad - 108 \cos A$$
$$25 = \quad\quad 117 \quad\quad - 108 \cos A$$
$$108 \cos A = 92$$
$$\angle A = \cos^{-1}\left(\frac{92}{108}\right)$$
$$= 31.6°, \text{ correct to the } \textit{nearest tenth of a degree}$$

Exercise Set 9.4

1. The resultant of a horizontal and a vertical force is a force of 200 pounds. If the resultant force makes an angle of 22° with the vertical force, and an angle of 38° with the horizontal force, find, to the *nearest pound,* the magnitude of the vertical force.

2. Two forces of 362 pounds and 529 pounds act upon a body at an acute angle with each other. The angle between the resultant and the 362-pound force is 35°40′. Find, to the *nearest 10 minutes,* the angle formed by the two given forces.

3. Two forces of 720 pounds and 640 pounds act upon a body at an angle of 42°40′ with each other. Find, to the *nearest 10 minutes,* the angle that the resultant force makes with the smaller force.

4. Two forces acting upon a body make an angle of 103°30′ with each other. The magnitude of the first force is 386 pounds. If the resultant makes an angle of 47°10′ with the first force, what is the magnitude of the resultant, to the nearest *tenth of a pound*?

5. Two forces of 70 pounds and 125 pounds act upon a body at an angle of 68° with each other. Find, to the *nearest 10 minutes,* the angle formed by the lines of action of the resultant and the larger force.

6. In Example 2, if $m\widehat{BC} = 170$, $m\widehat{BD} = 50$, $AD = 2$, and $CD = 6$, find the length of chord $\overline{BC}$.

7–8. In circle O, $\overleftrightarrow{ADB}$ is a secant and $\overleftrightarrow{BC}$ is tangent at C.

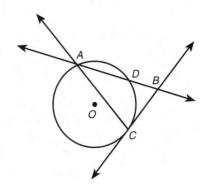

7. If $m\widehat{AD} : m\widehat{AC} : m\widehat{CD} = 2 : 7 : 3$ and $BC = 10$, find:
  (a) $\angle B$
  (b) AC correct to the *nearest tenth*

8. If $m\angle B = 45$, $BC = 12$, and $DB = 9$, find:
  (a) AB
  (b) AC correct to the *nearest tenth*
  (c) $\angle CAB$ correct to the *nearest degree*

Exercises 7 and 8

9 and 10. In circle O, $\overleftrightarrow{CDB}$ is a secant and $\overleftrightarrow{AB}$ is tangent at A.

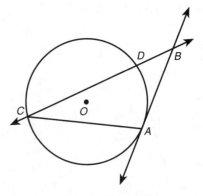

9. If $BD = 12$, $CD = 15$, and $m\widehat{AD} = 60$, find the measure of obtuse angle BAC correct to the *nearest 10 minutes*.

10. If $m\widehat{AD} = 60$, $m\widehat{CD} = 160$, $AB = 4$, and $BD = 2$, find:
  (a) CB
  (b) AC correct to the *nearest tenth*

Exercises 9 and 10

9.5 ANALYZING THE AMBIGUOUS CASE

═══════ △ KEY IDEAS △ ═══════

When the dimensions $a = 2$, $b = 5$, and $\sin A = 0.6$ are given, the value of $\sin B$ can be found by using the Law of Sines as follows:

$$\frac{a}{\sin A} = \frac{b}{\sin B} \qquad \text{or} \qquad a \sin B = b \sin A$$

$$2 \sin B = 5(0.6)$$

$$\sin B = \frac{3.0}{2} = 1.5$$

Since the sine of an angle cannot be greater than 1, a triangle is not possible. For other choices of a, b, and $\sin A$, however, it may be possible to construct one or two triangles. Since side-side-angle dimensions produce no, one, or two triangles, the situation in which SSA (Side-Side-Angle) is given is called the **ambiguous case.**

The situations in which the given angle is acute (Case I) and not acute (Case II) are considered separately.

The Ambiguous Case: Angle *A* Is Acute

If the measures a and b of two sides and an opposite angle, A, are given, as shown in Figure 9.4, then the altitude (height) h of a possible triangle may be found by using the sine ratio:

$$\sin A = \frac{h}{b} \qquad \text{or} \qquad h = b \sin A.$$

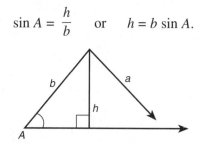

Figure 9.4 Ambiguous Case: ∠*A* Is Acute

A triangle can be constructed only if a, the side opposite the given acute angle, A, is at least as long as the altitude, h. The different situations are summarized in Tables 9.1 and 9.2.

TABLE 9.1 CASE I: ANGLE *A* IS ACUTE, AND THE SIDE OPPOSITE IS SHORTER THAN THE SIDE ADJACENT TO ∠*A*

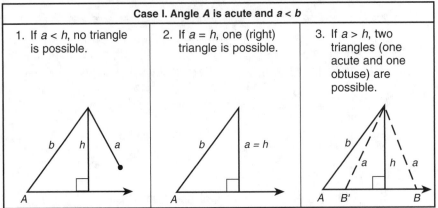

Case I. Angle *A* is acute and *a* < *b*		
1. If *a* < *h*, no triangle is possible.	2. If *a* = *h*, one (right) triangle is possible.	3. If *a* > *h*, two triangles (one acute and one obtuse) are possible.

TABLE 9.2 CASE I (CONTINUED): ANGLE *A* IS ACUTE AND THE SIDE OPPOSITE HAS THE SAME OR GREATER LENGTH THAN THE SIDE ADJACENT TO ∠*A*

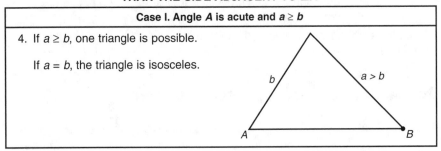

Case I. Angle *A* is acute and *a* ≥ *b*
4. If *a* ≥ *b*, one triangle is possible.
If *a* = *b*, the triangle is isosceles.

Examples

1. If *a* = 2, *b* = 5, and m∠*A* = 30, how many distinct triangles can be constructed?

Solution:
- Draw a diagram.
- Find the altitude, *h*.

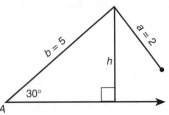

$$h = b \sin A$$
$$= 5 \sin 30°$$
$$= 5(0.5) = 2.5$$

- Compare *a* with *b* and *h*. Since *a* < *b* and *a* < *h*, **no** triangle can be constructed having the given dimensions.

2. In $\triangle ABC$, if $a = 7$, $b = 10$, and m$\angle A = 37$, then $\angle B$:
(1) is an acute angle   (3) is a right angle
(2) is an obtuse angle   (4) is either an acute or an obtuse angle

Solution:
- Draw a diagram.
- Find the altitude, h.

$$h = b \sin A$$
$$= 10 \sin 37°$$
$$\approx 10(0.6018) \approx 6.02$$

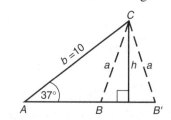

- Compare a with b and h. Since $a < b$ and $a > h$, two triangles can be constructed. Thus $\angle B$ may be either an acute or an obtuse angle. The correct choice is **(4)**.

3. If $a = 7$, $b = 10$, and m$\angle B = 50$, how many distinct triangles can be constructed?

Solution: Here the given angle is B, and the side opposite it is b. Since $b > a$, the length of the side opposite acute angle B is longer than the side adjacent to $\angle B$. Thus exactly **one** triangle can be constructed.

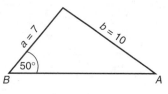

The Ambiguous Case: Angle *A* Is Not Acute

Table 9.3 summarizes the number of possible triangles that can be constructed, given SSA (Side-Side-Angle), when the given angle is either an obtuse or a right angle.

TABLE 9.3 CASE II: ANGLE *A* IS NOT ACUTE

Case II: Angle A is obtuse	
1. If $a \leq b$, no triangle is possible since the side opposite the greatest angle of a triangle must be its longest side.	2. If $a > b$, the side opposite the nonacute angle is longer than the adjacent side. Thus one triangle is possible.

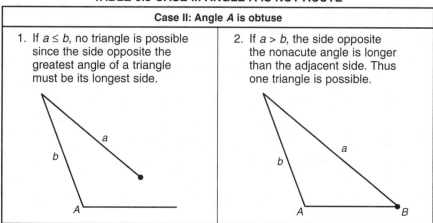

Example

4. If $r = 6$, $s = 5$, and m$\angle R = 100$, how many distinct triangles can be constructed?

Solution:
- Draw a diagram.
- Since the given angle is not acute, compare the side opposite with the side adjacent to the angle. Since r, the side opposite the obtuse angle, is longer than side s, **one** obtuse triangle can be constructed.

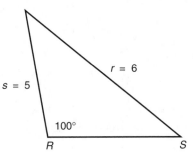

SUMMARY

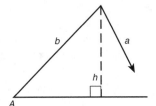

Case I:	$A < 90°$ and $a < b$			$A < 90°$
	$a < h$	$a = h$	$a > h$	$a \geq b$
Number of triangles:	0	1	2	1

Case II:	$A \geq 90°$	
	$a \leq b$	$a > b$
Number of triangles	0	1

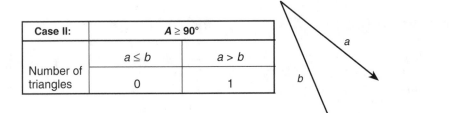

Exercise Set 9.5

1. If m$\angle ABC = 135$, $b = 9$, and $c = 10$, what is the maximum number of distinct triangles that can be constructed?
(1) 1   (2) 2   (3) 3   (4) 0

2. If m$\angle A = 32$, $a = 5$, and $b = 3$, it is possible to construct
(1) an obtuse triangle   (3) no triangles
(2) two distinct triangles   (4) a right triangle

3. Main Street and Central Avenue intersect, making an angle measuring 34°. Angela lives at the intersection of the two roads, and Caitlin lives on Central Avenue 10 miles from the intersection. If Leticia lives 7 miles from Caitlin, which conclusion is valid?
   (1) Leticia cannot live on Main Street.
   (2) Leticia can live at only one location on Main Street.
   (3) Leticia can live at one of two locations on Main Street.
   (4) Leticia can live at one of three locations on Main Street.

4. If $\sin A = 0.75$ and $b = 8$, for which value of a can two distinct triangles be constructed?
   (1) 5   (2) 6   (3) 7   (4) 9

5. If $a = 5$ and $b = 10$, for which measure of $\angle A$ can two distinct triangles be constructed?
   (1) 20   (2) 30   (3) 45   (4) 120

6–8. In each case use the given data to determine the number of distinct triangles that can be constructed.

6. Given: $m\angle R = 100$, $r = 7$ and $s = 5$.

7. Given: $m\angle A = 43$, $a = 15$ and $b = 24$.

8. Given: $r = 7$, $s = 6$, and $m\angle R = 50$.

9–13. In each case, use the given data to determine whether it is possible to construct no triangle, two triangles, a right triangle, or an obtuse triangle.

9. Given: $\angle S = 100°$, $s = 8$, and $t = 9$.

10. Given: $\angle A = 34°20'$, $a = 57.0$, and $b = 100.0$.

11. Given: $\angle A = 40°10'$, $a = 12$, and $b = 20$.

12. Given: $\angle T = 30°$, $t = 4.6$, and $s = 9.2$.

13. Given: $\angle A = 60$, $a = 5\sqrt{2}$, and $b = 8$.

Each of the questions in this section has appeared on a previous Course III Regents Examination. Here is an opportunity for you to review Chapter 9 and, at the same time, prepare for the Course III Regents Examination.

1. In $\triangle ABC$, $a = 5$, $\sin A = \dfrac{1}{5}$, and $b = 4$. Find the value of $\sin B$.

2. In $\triangle ABC$, $b = 3$, $c = 4$, and $\angle A = 45°$. Expressed in simplest radical form, what is the area of $\triangle ABC$?

3. In $\triangle ABC$, $a = 2$, $b = 3$, and $c = 4$. What is the value of $\cos C$?

4. In $\triangle ABC$, if $\sin A = \dfrac{4}{5}$, $\sin B = \dfrac{3}{8}$, and $a = 24$, find b.

5. In $\triangle ABC$, $a = \sqrt{3}$, $b = \sqrt{3}$, and $m\angle C = 120$. Find C.

6. If $m\angle A = 30$, $a = 11$, and $b = 12$, the number of distinct triangles that can be constructed is:
   (1) 1   (2) 2   (3) 3   (4) 0

7. If $a = 5\sqrt{2}$, $b = 8$, and $m\angle A = 45$, how many distinct triangles can be constructed?
   (1) 1   (2) 2   (3) 3   (4) 0

8. The area of $\triangle ABC$ is 20. If $a = 10$ and $b = 8$, find the number of degrees in the measure of acute angle C.

9. If $m\angle A = 48$, $a = 7$, and $b = 9$, the number of distinct triangles that can be constructed is:
   (1) 1   (2) 2   (3) 3   (4) 0

10. The lengths of the sides of a triangular plot of land are 50, 80, and 100 meters.
    (a) Find, to the *nearest degree,* the measure of the *largest* angle of the triangle.
    (b) Using the answer obtained in part (a), find, to the *nearest square meter,* the area of the triangle.

11. Two forces of 30 pounds and 40 pounds act upon a body, forming an acute angle with each other. The angle between the resultant and the 30-pound force is $35°10'$. Find, to the *nearest 10 minutes,* the angle between the two given forces.

12. In the accompanying diagram $\overleftrightarrow{DCFG}$ is tangent to circle O at F, $\overleftrightarrow{ECH}$ is a secant intersecting the circle at A and B, $m\overarc{AB} = 140$, and $m\overarc{AF} = 160$. If $FC = 10$, find AF to the *nearest tenth*.

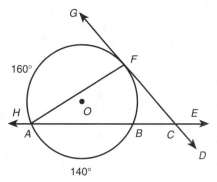

160°

140°

13. In the accompanying diagram, tangent $\overline{PB}$ and secant $\overleftrightarrow{PCA}$ are drawn to circle O, $AB = 8\sqrt{2}$, and $m\overarc{BC} : m\overarc{AC} : m\overarc{AB} = 3 : 4 : 5$.
   (a) Find the measure of:
      (1) $\overarc{BC}$ (2) $\angle PAB$ (3) $\angle APB$
   (b) Find the length of $\overline{PB}$.

14. The building lot shown in the accompanying diagram is shaped like an isosceles triangle with $AB = AC$ and $\angle BAC = 53°10'$. The area of the lot is 1 acre. Find the length, to the *nearest foot,* of each of the three sides. [1 acre = 43,560 square feet]

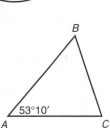

53°10'

15. To determine the distance across a river, a surveyor marked three points on one riverbank: H, G, and F, as shown below. She also marked one point, K, on the opposite bank such that $\overline{KH} \perp \overline{HGF}$, m$\angle KGH = 41$, and m$\angle KFH = 37$. The distance between G and F is 45 meters. Find KH, the width of the river, to the *nearest tenth of a meter*.

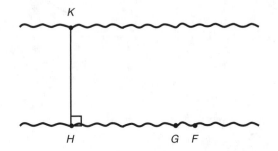

16. Two forces of 130 and 150 pounds yield a resultant force of 170 pounds. Find, to the *nearest ten minutes* or *nearest tenth of a degree*, the angle between the original two forces.

17. Two forces are applied to an object. The measure of the angle between the 30.2-pound applied force and the 50.1-pound resultant is 25°.
   (a) Find, to the *nearest tenth of a pound*, the magnitude of the second applied force.
   (b) Using the answer found in part **(a)**, find, to the *nearest degree*, the measure of the angle between the second applied force and the resultant.

ANSWERS TO SELECTED EXERCISES: CHAPTER 9

Section 9.1

1. (a) 21	**(c)** $21\sqrt{3}$	**3.** 4.2	**7.** 8
(b) $21\sqrt{2}$	**(d)** 214.	**4.** 0.075	**8.** 8
2. (a) 47	**(c)** 98	**5.** $30\sqrt{3}$	**9.** 130.2°
(b) 91	**(d)** 88	**6.** 5.8	

Section 9.2

1. 0.54	**4.** $3\sqrt{2}$	**7.** 46	**10.** 4240
2. 9	**5.** $\dfrac{1}{4}$	**8.** 48.5	**11.** 16.71
3. 0.54	**6.** 473	**9.** 68°20′	

Section 9.3

1. 7

2. (3)

3. $\frac{1}{7}$

4. (2)

5. $\frac{13}{14}$

6. 120°

7. $\frac{61 - c^2}{60}$

8. $3\sqrt{3}$

9. (a) 109°30′
   (b) 2515

10. 36

11. (a) 101°30′   (b) 34°0′

12. 54.8

13. 10.6

14. (a) 97.9°   (b) 3962

15. (a) 104°   (b) 159

Section 9.4

1. 142

2. 59°10′

3. 25°40′

4. 451.0

5. 23°10′

6. $4\sqrt{3}$

7. (a) 60°   (b) 11.1

8. (a) 16   (b) 25.9   (c) 19°

9. 131°20′

10. (a) 8   (b) 5.1

Section 9.5

1. (4)

2. (1)

3. (3)

4. (3)

5. (4)

6. 1

7. 0

8. 1

9. No triangle

10. Two triangles

11. No triangle

12. A right triangle

13. No triangle

Regents Tune-Up: Chapter 9

1. $\frac{4}{25}$

2. $3\sqrt{2}$

3. −0.25

4. 11.25

5. 3

6. (2)

7. (2)

8. 30

9. (2)

10. (a) 98°   (b) 1981

11. 60°50′

12. 15.3

13. (a)   (1) 90°
        (2) 45°
        (3) 30°
   (b)   16

14. $AB = AC \approx 330$, $BC \approx 295$

15. 254.7

16. 105.40′ or 105.6°

17. (a) 26.1   (b) 29°

PROBABILITY AND STATISTICS

CHAPTER 10

BERNOULLI EXPERIMENTS, THE BINOMIAL THEOREM, AND THE NORMAL CURVE

10.1 FINDING PROBABILITIES IN TWO-OUTCOME EXPERIMENTS

KEY IDEAS

In tossing a coin there are two possible outcomes, head and tail. If a probability experiment that has exactly two possible outcomes is repeated, as in tossing a coin five times, it is called a **Bernoulli experiment**.

Each repetition of a Bernoulli experiment is called a *trial.* If a coin is flipped five times, then there are five trials. For each trial the concern is whether a particular outcome occurs or does not occur. If the probability of one of the outcomes occurring in a single trial is known, the probability that this outcome will occur *r* times in *n* trials can be easily determined.

Characteristics of a Bernoulli Experiment

A Bernoulli experiment has n identical trials such that all of the following are true:

- Each trial results in one of two possible outcomes. One of these outcomes is considered a "success," while the other is labeled a "failure."
- The n trials are *independent*. The outcome of one trial has no effect on the outcome of any other trial.
- The probability of a success in a single trial is usually denoted by the letter p, while the probability of a failure is represented by the letter q. Since the sum of the probabilities of the possible outcomes of any experiment is 1,

$$p + q = 1 \quad \text{or} \quad q = 1 - p.$$

- If there are r successes in n trials $(r \leq n)$, there must be $n - r$ failures.

Obtaining *r* Successes, Followed by All Failures

Suppose a coin is weighted so that the probability of obtaining a head is $\frac{2}{3}$. If

$$p = P(\text{head}) = \frac{2}{3}, \quad \text{then} \quad q = P(\text{tail}) = 1 - p = \frac{1}{3}.$$

If this coin is tossed five times, the probability of obtaining a head on the first toss *followed by* four tails on subsequent tosses is calculated by multiplying the probability of the desired outcome occurring (head or tail) on each of the five trials:

$$\frac{2}{3} \cdot \frac{1}{3} \cdot \frac{1}{3} \cdot \frac{1}{3} \cdot \frac{1}{3} = \left(\frac{2}{3}\right)^1 \cdot \left(\frac{1}{3}\right)^4.$$

The probability of obtaining a head on each of the first two tosses, *followed by* three tails, is

$$\frac{2}{3} \cdot \frac{2}{3} \cdot \frac{1}{3} \cdot \frac{1}{3} \cdot \frac{1}{3} = \left(\frac{2}{3}\right)^2 \cdot \left(\frac{1}{3}\right)^3.$$

Similarly, the probability of obtaining a head on each of the first three tosses, *followed by* two tails, is

$$\frac{2}{3} \cdot \frac{2}{3} \cdot \frac{2}{3} \cdot \frac{1}{3} \cdot \frac{1}{3} = \left(\frac{2}{3}\right)^3 \cdot \left(\frac{1}{3}\right)^2.$$

Notice that on the right side of each of the preceding equations the power of $\frac{2}{3} \, (= p)$ represents the number of successes, the power of $\frac{1}{3} \, (= q)$ represents the number of failures, and the sum of the exponents equals the number of trials.

=== **MATH FACTS** ===

BERNOULLI EXPERIMENT WITH *r* SUCCESSES FIRST

$P(r \text{ successes}, \ n - r \text{ failures}) = p^r q^{n-r}$,
where $p + q = 1$.

Combination Notation

The number of different ways in which *r* things can be selected from a group of *n* things $(r \le n)$, without paying attention to the order in which the *r* things are selected, is represented by $_nC_r$. The notation $_nC_r$ is read as "the combination of *n* things taken *r* at a time," and can be evaluated using the formula

$$_nC_r = \frac{_nP_r}{r!} = \frac{n!}{r!(n-r)!}.$$

For example, to evaluate $_5C_3$, let $n = 5$ and $r = 3$; then

$$_5C_3 = \frac{5!}{3!(5-3)!} = \frac{5 \cdot 4 \cdot 3 \cdot 2 \cdot 1}{(3 \cdot 2 \cdot 1)(2 \cdot 1)} = \mathbf{10}.$$

Instead of using a complicated formula to figure out the number of combinations, you can use your scientific calculator. Since $_nC_r$ is usually printed above another calculator key, you will probably need to press the ⟨2nd⟩ function key before pressing the key that has $_nC_r$ printed above it. For example, to evaluate $_5C_3$, locate $_nC_r$ on your calculator and then key in

5　⟨2nd⟩　⟨$_nC_r$⟩　3　⟨=⟩.

The display window now shows 10 since $_5C_3 = 10$.

MATH FACTS

EVALUATING COMBINATIONS

To evaluate $_nC_r$ for given values of n and r, key in:

n　⟨2nd⟩　⟨$_nC_r$⟩　r　⟨=⟩.

If you remember the following relationships, you can often save time in evaluating combinations having the indicated form:

- $_nC_n = 1$ 　　Example: $_5C_5 = 1$
- $_nC_0 = 1$ 　　Example: $_5C_0 = 1$
- $_nC_1 = n$ 　　Example: $_5C_1 = 5$

Obtaining r Successes in n Trials ($r \le n$)

The r successes of a Bernoulli experiment need not occur consecutively in the first r trials. Within the n trials, the r successes ($r \le n$) can be arranged in $_nC_r$ ways. Multiplying $_nC_r$ by $p^r q^{n-r}$ gives the probability of obtaining exactly r successes in any order within the n trials of a Bernoulli experiment.

For example, if the probability of winning a game is .7, then the probability of winning 3 out of 4 games is found by using the formula $_nC_r p^r q^{n-r}$, where $p = .7$, $q = .3$, $r = 3$, and $n = 4$:

$$P(\text{win } r \text{ out of } n \text{ games}) = {_nC_r}p^r q^{n-r}$$
$$P(\text{win 3 out of 4 games}) = {_4C_3}\,(.7)^3(.3)^{4-3}$$
$$= 4(.343)(.3)$$
$$= \mathbf{.4116}$$

BERNOULLI EXPERIMENT WITH r SUCCESSES IN n TRIALS

$P(r$ successes in n Bernoulli trials$) = {}_nC_r p^r q^{n-r}$,

where

$$ {}_nC_r = \frac{n!}{r!(n-r)!} \quad \text{and} \quad p + q = 1. $$

Example

In a certain part of the country, the probability of rain on any day is $\frac{1}{5}$. In the next 5 days, what is the probability that in that region it will rain: **(a)** *exactly* 3 days **(b)** *at least* 3 days **(c)** *at most* 3 days

Solutions: This is a Bernoulli experiment in which there are two possible outcomes: (1) it rains or (2) it does not rain. Let $p = \frac{1}{5}$, $q = \frac{4}{5}$, and $n = 5$.

(a) Use the formula ${}_nC_r p^r q^{n-r}$, where $r = 3$.

$$ P(\text{rains } \textit{exactly} \text{ 3 days out of 5}) = {}_5C_3 \left(\frac{1}{5}\right)^3 \left(\frac{4}{5}\right)^2 $$

$$ = 10\left(\frac{1}{125}\right)\left(\frac{16}{25}\right) $$

$$ = \frac{160}{3125} $$

(b) If it rains on *at least* 3 days, then it may rain on 3, 4, or 5 days. Hence the probability of rain on at least 3 days is the sum of the probabilities of rain on 3 days, 4 days, and 5 days.

$$ P(\text{rains 3 days out of 5}) = \frac{160}{3125} $$

$$ P(\text{rains 4 days out of 5}) = {}_5C_4 \left(\frac{1}{5}\right)^4 \left(\frac{4}{5}\right)^1 $$

$$ = 5\left(\frac{1}{625}\right)\left(\frac{4}{5}\right) $$

$$ = \frac{20}{3125} $$

$$P(\text{rains 5 days out of 5}) = {}_5C_5\left(\frac{1}{5}\right)^5\left(\frac{4}{5}\right)^0$$

$$= 1\left(\frac{1}{3125}\right)1$$

$$= \frac{1}{3125}$$

Adding these probabilities gives the probability of rain on *at least* 3 days:

$$P(\text{rains } at \ least \ 3 \text{ days out of 5}) = \frac{160}{3125} + \frac{20}{3125} + \frac{1}{3125} = \frac{181}{3125}.$$

(**c**) If it rains on *at most* 3 days, then it may rain on 0, 1, 2, or 3 days. Hence the probability of rain on *at most* 3 days is the sum of the probability of rain on 0, 1, 2, and 3 days.

$$P(\text{rains 0 day out of 5}) = {}_5C_0\left(\frac{1}{5}\right)^0\left(\frac{4}{5}\right)^5 = \frac{1024}{3125}$$

$$P(\text{rains 1 day out of 5}) = {}_5C_1\left(\frac{1}{5}\right)^1\left(\frac{4}{5}\right)^4 = \frac{1280}{3125}$$

$$P(\text{rains 2 days out of 5}) = {}_5C_2\left(\frac{1}{5}\right)^2\left(\frac{4}{5}\right)^3 = \frac{640}{3125}$$

$$P(\text{rains 3 days out of 5}) = {}_5C_3\left(\frac{1}{5}\right)^3\left(\frac{4}{5}\right)^2 = \frac{160}{3125}$$

Adding these probabilities gives the probability of rain on *at most* 3 days:

$$P(\text{rains } at \ most \ 3 \text{ days}) = \frac{1024}{3125} + \frac{1280}{3125} + \frac{640}{3125} + \frac{160}{3125} = \frac{3104}{3125}.$$

Exercise Set 10.1

1. The probability of winning a game is $\frac{3}{5}$, and the probability of losing a game is $\frac{2}{5}$. If the game is played three times, what is the probability of winning *exactly* two games?

2. If the probability that Mike will successfully complete a foul shot is $\frac{4}{5}$, what is the probability that he will successfully complete *exactly* three of his next four foul shots?

(1) $\dfrac{64}{625}$ (2) $\dfrac{192}{625}$ (3) $\dfrac{256}{625}$ (4) $\dfrac{64}{125}$

3. The probability of Rick getting an A on any test is $\frac{2}{5}$. Which expression represents the probability that he will earn an A on *exactly* three of four tests?

(1) $_5C_4\left(\dfrac{2}{5}\right)^4\left(\dfrac{3}{5}\right)$ (3) $_5C_4\left(\dfrac{3}{5}\right)^4\left(\dfrac{2}{5}\right)$

(2) $_4C_3\left(\dfrac{2}{5}\right)^3\left(\dfrac{3}{5}\right)$ (4) $_4C_3\left(\dfrac{3}{5}\right)^3\left(\dfrac{2}{5}\right)$

4. The probability of Gordon's team winning any given game in a five-game series is 30%. Which expression represents the probability that Gordon's team will win *exactly* two games in the series?

(1) $(0.3)^2(0.7)^3$ (3) $10(0.3)^2(0.7)^3$
(2) $5(0.3)^3(0.7)^2$ (4) $5(0.3)^2(0.7)$

5. A box contains eight good and four bad lightbulbs. Daniel randomly picks a bulb from the box, tests it, and then replaces it. If Daniel picks four lightbulbs with replacement, what is the probability he will pick *exactly* two bad lightbulbs?

(1) $\dfrac{4}{27}$ (2) $\dfrac{1}{9}$ (3) $\dfrac{8}{27}$ (4) $\dfrac{4}{9}$

6. Each day the probability of rain on a tropical island is $\frac{7}{8}$. Which expression represents the probability that there will be rain on the island *exactly x* days in the next 3 days?

(1) $_3C_x\left(\dfrac{7}{8}\right)^x\left(\dfrac{1}{8}\right)^{3-x}$ (3) $_xC_3\left(\dfrac{7}{8}\right)^3\left(\dfrac{1}{8}\right)^x$

(2) $_3C_3\left(\dfrac{7}{8}\right)^3\left(\dfrac{1}{8}\right)^x$ (4) $_xC_3\left(\dfrac{7}{8}\right)^{3-x}\left(\dfrac{1}{8}\right)^x$

7. At a certain intersection, the light for eastbound traffic is red for 15 seconds, yellow for 5 seconds, and green for 30 seconds. Find, to the *nearest tenth*, the probability that, of the next eight eastbound cars that arrive randomly at the light, *exactly* three will be stopped by a red light.

8. Team A and team B are playing in a league. They will play each other five times. If the probability that team A wins a game is $\dfrac{1}{3}$, what is the probability that team A will win *at least* three of the five games?

9. Circle O is divided into <u>four</u> regions, as shown in the accompanying diagram, with diameter $\overline{ROS}$. A fair spinner is placed at center O and spun three times. Find the probability that the spinner will stop in:
   (a) region A or B all three times
   (b) region D *at most* one time

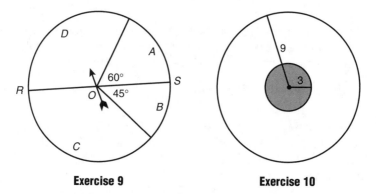

| Exercise 9 | Exercise 10 |

10. As shown in the accompanying diagram, a circular target with a radius of 9 inches has a bull's-eye that has a radius of 3 inches. If five arrows randomly hit the target, what is the probability that *at least* four hit the bull's-eye?

11. The probability that a new calculator battery will fail within the first 100 hours of use is 30%.
   (a) A graphing calculator will power on only if none of its four batteries fails. If four new batteries are put into a graphing calculator, what is the probability that the calculator will *always* power on within the next 100 hours of calculator use?
   (b) What is the probability that *no more than one* of the original four batteries will fail within the first 100 hours of calculator use?

10.2 EXPANDING $(a + b)^n$

========= ∧ =========
KEY IDEAS
∠ ＼

A binomial such as $(a + b)^5$ may be expanded by repeated multiplication, using $(a + b)$ as a factor five times. An algebraic formula called the **binomial theorem** provides a simpler way of finding the same product.

Binomial Theorem

Using repeated multiplication, you can find that

$$(a+b)^5 = 1 \cdot a^5 + 5a^4b^1 + 10a^3b^2 + 10a^2b^3 + 5a^1b^4 + 1 \cdot b^5.$$

Notice that:

- The expansion of $(a + b)^5$ has $5 + 1$ or 6 terms. The first term is a^5, and the last term is b^5.
- The terms of the expansion contain the products of descending powers of a and ascending powers of b so that the sum of the exponents of each term is 5.
- The numerical coefficients of the terms of the expansion can be represented using combination notation, as shown in Table 10.1.

TABLE 10.1 REPRESENTATIONS OF THE NUMERICAL COEFFICIENTS OF $(a + b)^5$

Term Number	Numerical Coefficient
1	$1 = {}_5C_0$
2	$5 = {}_5C_1$
3	$10 = {}_5C_2$
4	$10 = {}_5C_3$
5	$5 = {}_5C_4$
6	$1 = {}_5C_5$

From Table 10.1, you can generalize that the numerical coefficient of the kth term of $(a + b)^n$ is ${}_nC_{k-1}$. For example, the numerical coefficient of the fourth term of $(a + b)^5$ can be obtained by evaluating ${}_5C_{k-1}$ for $k = 4$, which gives ${}_5C_{4-1} = {}_5C_3 = 10$. Similarly, to find the numerical coefficient of the third term of $(a + b)^7$, evaluate ${}_7C_{k-1}$ for $k = 3$, which gives ${}_7C_{3-1} = {}_7C_2 = 21$.

===== **MATH FACTS** =====

BINOMIAL THEOREM

When a binomial of the form $(a + b)^n$ is expanded, the terms follow a pattern that is summarized by the binomial theorem:

$$(a + b)^n = {}_nC_0 \cdot a^n + {}_nC_1 \cdot a^{n-1}b^1 + {}_nC_2 \cdot a^{n-2}b^2 + {}_nC_3 \cdot a^{n-3}b^3 + \ldots + {}_nC_{n-1} \cdot a^1 b^{n-1} + {}_nC_n \cdot b^n.$$

The numerical coefficient of the first term is represented by ${}_nC_0$. Since the expansion of $(a + b)^n$ has $n + 1$ terms, the numerical coefficient of the kth term is represented by ${}_nC_{k-1}$.

Example 1 shows how the binomial theorem can be used to expand a power of a binomial of the form $(a + b)^n$ by substituting values into a formula, rather than performing repeated multiplications of the binomial.

Example

1. Expand $(3x - y)^4$ using the binomial theorem.

Solution: For $n = 4$, the binomial theorem states:

$$(a + b)^4 = {}_4C_0 a^4 + {}_4C_1 a^3 b + {}_4C_2 a^2 b^2 + {}_4C_3 ab^3 + {}_4C_4 b^4.$$

Let $a = 3x$ and $b = -y$. Then

$$(3x - y)^4 = {}_4C_0(3x)^4 + {}_4C_1(3x)^3(-y) + {}_4C_2(3x)^2(-y)^2$$
$$+ {}_4C_3(3x)(-y)^3 + {}_4C_4(-y)^4$$

Simplify, and substitute ${}_4C_0 = {}_4C_4 = 1$, ${}_4C_1 = 4$, ${}_4C_2 = 6$, and ${}_4C_3 = 4$:

$$(3x - y)^4 = (1)(81x^4) + (4)(27x^3)(-y) + (6)(9x^2)(y^2)$$
$$+ (4)(3x)(-y^3) + (1)(y^4)$$
$$= 81x^4 - 108x^3y + 54x^2y^2 - 12xy^3 + y^4$$

Writing the *k*th Term

Sometimes the only thing you want to know about the power of a binomial is what a particular term looks like. Specific terms in a binomial expansion of the form $(a + b)^n$ are identified by their position numbers in the series. For example, since the expansion of $(a + b)^5$ is as follows:

| First term | Second term | Third term | Fourth term | Fifth term | Sixth term |

$$_5C_0a^5 \; + \; _5C_1a^4b \; + \; _5C_2a^3b^2 \; + \; _5C_3a^2b^3 \; + \; _5C_4ab^4 \; + \; _5C_5b^5$$

the third term is $_5C_2a^3b^2$. In the expansion of $(a + b)^n$:

- The numerical coefficient of the kth term is $_nC_{k-1}$.
- The exponent of b in the kth term is $k - 1$, and the exponent of a is $n - (k - 1)$. For example, in the **third** term the exponent of b is $3 - 1 = 2$, and the exponent of a is $5 - (3 - 1) = 5 - 2 = 3$.

MATH FACTS

kTH TERM OF $(a + b)^n$

$$k\text{th term} = {}_nC_{k-1}a^{n-(k-1)}b^{k-1}$$

For example, the second term in the expansion of $(3x - y)^4$ may be found by letting $a = 3x$, $b = -y$, $n = 4$, and $k = 2$:

$$\begin{aligned} \text{second term} &= {}_4C_{2-1}(3x)^{4-(2-1)}(-y)^{2-1} \\ &= (4)(3x)^3(-y)^1 \\ &= (4)(27x^3)(-y) \\ &= \mathbf{-108x^3y} \end{aligned}$$

Examples

2. What is the middle term in the expansion of $(x - 2y)^4$?

Solution: Since the expansion of $(a + b)^n$ consists of $n + 1$ terms, the expansion of $(x - 2y)^4$ has $4 + 1$ or 5 terms. The middle term is the third term. To find the third term let $a = x$, $b = -2y$, $n = 4$, and $k = 3$:

$$\begin{aligned} {}_nC_{k-1}a^{n-(k-1)}b^{k-1} &= {}_4C_{3-1}x^{4-(3-1)}(-2y)^{3-1} \\ &= {}_4C_2x^2(-2y)^2 \\ &= 6x^2(4y^2) = \mathbf{24x^2y^2} \end{aligned}$$

3. What is the numerical coefficient of the term that contains x^5y^4 in the expansion of $(x + yi)^9$, where $i = \sqrt{-1}$?

Solution: The first term contains x^9, the second term contains x^8y, the third term contains x^7y^2, the fourth term contains x^6y^3, and the *fifth* term contains x^5y^4. Let $a = x$, $b = yi$, $n = 9$, and $k = 4$:

$$\begin{aligned} {}_nC_{k-1}a^{n-(k-1)}b^{k-1} &= {}_9C_{5-1}x^{9-(5-1)}(yi)^{5-1} \\ &= 126x^5y^4i^4 \end{aligned}$$

Since $i^4 = 1$, the fifth term is $126x^5y^4$. The numerical coefficient of the term that contains x^5y^4 is **126**.

Exercise Set 10.2

1 and 2. Use the binomial theorem to write each expansion.

1. $(x + y)^4$ **2.** $(p - 3q)^4$

3. What is the fourth term in the expansion of $(2x - y)^7$?
   (1) $16x^4y^3$ (2) $35x^3y^4$ (3) $-560x^4y^3$ (4) $-560x^3y^4$

4. What is the third term in the expansion of $(\sin x - \cos y)^5$?
   (1) $10 \sin^3 x \cos^2 y$ (3) $10 \sin^2 x \cos^3 y$
   (2) $-10 \sin^3 x \cos^2 y$ (4) $-10 \sin^2 x \cos^3 y$

5. The fifth term in the expansion of $(3a - b)^6$ is
   (1) $135a^2b^4$ (2) $540a^3b^3$ (3) $-18ab^5$ (4) $-135a^2b^4$

6. The third term of the expansion of $(\frac{1}{2}x - 2y)^4$ is
   (1) $6x^2y^2$ (2) $-6x^2y^2$ (3) $16xy^3$ (4) $-16xy^3$

7 and 8. Find the third term in each expansion.

7. $(p + q)^8$ **8.** $(3x - 5y)^6$

9 and 10. Find the fourth term in each expansion.

9. $(x - 2)^7$ **10.** $\left(a - \dfrac{b}{2}\right)^6$

11 and 12. Find the fifth term in each expansion.

11. $(r - 2s)^5$ **12.** $(a + bi)^7$ [where $i = \sqrt{-1}$]

13. What is the numerical coefficient of the term x^4y^2 in the expansion of $(x + y)^6$?

14. What is the numerical coefficient of the term c^5 in the expansion of:

   (a) $(b - 2c)^5$ **(b)** $\left(b + \dfrac{c}{3}\right)^6$

15 and 16. In the expansion of each of the following binomials, find:

   (a) the middle term **(b)** the next-to-last term

15. $\left(10 - \dfrac{x}{2}\right)^4$ **16.** $(a - bi)^{10}$ [where $i = \sqrt{-1}$]

10.3 COMPARING STATISTICAL MEASURES

$\overset{\wedge}{\underset{\diagup\diagdown}{\text{KEY IDEAS}}}$

Much of statistics is concerned with finding ways to describe how the individual numbers in a set are distributed in value.

The *mean, median,* and *mode* are measures of **central tendency** since they indicate how the data values are grouped about some central value.

The *range, mean, absolute deviation, variance,* and *standard deviation* are measures of **dispersion** since they indicate whether the data are spread out or are clustered together.

Since no single statistical measure describes a set of data values completely, when analyzing data you need to look at measures of dispersion as well as measures of central tendency.

Statistics

Statistics is the branch of mathematics that deals with collecting, organizing, displaying, and interpreting data. Data are facts and figures. Numerical data may be referred to as values, scores, measures, or observations.

Subscripted Variables

Sometimes it is convenient to use a single variable, say x, to represent an entire list of data values. An individual data value in the list may be named by writing x followed by a number, called a subscript, that represents its position in the list. For example, the first data value in a list is called x_1 (read as "x sub 1"), the second data value is called x_2, and so forth. Notice that the subscript is written to the right of the variable, one-half line down from it.

Variables such as x_1 and x_2 are called *subscripted variables*. If $x = \{15, 7, 4, 13, 11\}$, then

$$x_1 = 15, \quad x_2 = 7, \quad x_3 = 4, \quad x_4 = 13, \quad \text{and} \quad x_5 = 11.$$

Sigma Notation

Many of the formulas used in statistics involve finding the sum of a series of numbers. The symbol Σ is the capital Greek letter *sigma* and indicates that the sum of whatever follows should be taken. For example, the notation

$\displaystyle\sum_{i=1}^{5} x_i$ represents the sum of the values of x_i as i increases from 1 to 5, in steps of 1:

$$\sum_{i=1}^{5} x_i = x_1 + x_2 + x_3 + x_4 + x_5$$
$$= 15 + 7 + 4 + 13 + 11 = 50$$

In the summation $\sum_{i=1}^{n} x_i$, i is called the *index variable*, 1 is the *lower limit* (first value for i), and n is the *upper limit* (last value for i).

Here are some additional facts about summation notation that you should know:

- The index variable may start at a whole number different from 1. For example,

$$\sum_{i=3}^{5} x_i = x_3 + x_4 + x_5.$$

- If a constant factor appears inside the summation sign, it may be "passed through it." For example,

$$\sum_{i=1}^{5} 2x_i = 2 \sum_{i=1}^{5} x_i$$
$$= 2(x_1 + x_2 + x_3 + x_4 + x_5)$$

- The index variable may be used to represent the terms to be added. For example,

$$\sum_{i=1}^{5} i^2 = 1^2 + 2^2 + 3^2 + 4^2 + 5^2$$
$$= 1 + 4 + 9 + 16 + 25$$
$$= 55$$

Example

1. Evaluate:

(a) $\sum_{k=1}^{3} (2k-1)^2$ **(b)** $\sum_{n=3}^{4} \frac{1}{7}(2^n)$ **(c)** $\sum_{p=1}^{3} 64^{\frac{1}{p}}$

Solutions: **(a)** The summation symbol tells you to evaluate and then accumulate terms of the form $(2k-1)^2$ for $k = 1, 2,$ and 3.

$$\sum_{k=1}^{3} (2k-1)^2 = (2 \cdot 1 - 1)^2 + (2 \cdot 2 - 1)^2 + (2 \cdot 3 - 1)^2$$

$$
\begin{aligned}
&= (2-1)^2 &&+ (4-1)^2 &&+ (6-1)^2 \\
&= 1^2 &&+ 3^2 &&+ 5^2 \\
&= 1 &&+ 9 &&+ 25 \\
&= \mathbf{35}
\end{aligned}
$$

(b) Since the starting value of the index variable is 3, the summation symbol tells you to evaluate and then accumulate terms of the form $\frac{1}{7}(2^n)$ for $n = 3, 4$, and 5.

$$\sum_{n=3}^{5} \frac{1}{7}(2^n) = \frac{1}{7} \sum_{n=3}^{5} 2^n$$

$$= \frac{1}{7}(2^3 + 2^4 + 2^5)$$

$$= \frac{1}{7}(8 + 16 + 32)$$

$$= \frac{1}{7}(56) = \mathbf{8}$$

(c) $\sum_{p=1}^{3} 64^{\frac{1}{p}} = 64^{\frac{1}{1}} + 64^{\frac{1}{2}} + 64^{\frac{1}{3}}$

$$= 64 + 8 + 4$$
$$= \mathbf{76}$$

Mean, Median, and Mode

The **mean** (average) of a set of data values is the sum of the values divided by the number of values. Thus, if subscripted variables of the form x_i represent the individual data values, $\bar{x}$ (read as "x bar") is commonly used to represent the mean.

If there are n data values, then

$$\text{Mean} = \bar{x} = \frac{1}{n} \sum_{i=1}^{n} x_i = \frac{1}{n}(x_1 + x_2 + x_3 + \cdots + x_n).$$

For example, if $x = \{15, 7, 4, 13, 1\}$, then

$$\bar{x} = \frac{1}{5}(15 + 7 + 4 + 13 + 1) = \frac{1}{5}(50) = \mathbf{10}.$$

The **median** of a set of data values is the middle value when the numbers are arranged in size order. To find the median of $\{15, 7, 4, 13, 1\}$, arrange the numbers is ascending (or descending) order: $\{1, 4, 7, 13, 15\}$. The median is **7**, since 7 is the middle value of the set.

When an ordered data set has an even number of data values, the median is found by taking the average of the two middle values. The median of $\{6, 17, 18, 22, 23, 31\}$ is **20**, since 20 is the average of 18 and 22.

The **mode** is the most frequently occurring data value in a set of observations. The mode of $\{2, 3, 1, 2, 4, 3, 5, 6, 2, 5\}$ is **2**, since 2 appears more times than any other value in the list.

Measures of Dispersion

Consider the two sets of numbers:

$$A = \{49, 53, 51, 55\} \quad \text{and} \quad B = \{1, 2, 5, 200\}.$$

The mean of each set is 52. For set A, the mean of 52 is fairly representative of the members of the set. The same is not true, however, for set B since each of its measures varies greatly from the mean of 52. Since it is useful to know whether the mean is a representative data value, additional statistical measures must be introduced that indicate the extent to which the individual data values are scattered or dispersed about the mean. Statistics that measure this type of variability are called **measures of dispersion**.

The simplest measure of dispersion is the *range*. The **range** of a set of numbers is the difference between the largest and the smallest numbers of the set. The range of the set 1, 2, 5, 200 is 199 since $200 - 1 = 199$. When the range is relatively large, the value of the mean *may* be distorted.

Standard Deviation

The **standard deviation** is the most widely used measure of dispersion. It indicates how widely scattered the values in a set of data are about the mean. The letter s or σ (the lower-case Greek letter sigma) is commonly used to represent the standard deviation. Sometimes it is abbreviated as S.D.

Although the standard deviation σ of a set of n data values can be calculated using the formula

$$\sigma = \sqrt{\frac{1}{n} \sum_{i=1}^{n} (x_i - \bar{x})^2},$$

the arithmetic may be lengthy and error-prone. After a set of data values has been entered and stored in your scientific calculator, the mean and standard deviation of those values can be easily obtained by pressing a couple of keys. If you are using a Casio scientific calculator, you will need to have the calculator in the SD (Statistical *D*ata) mode before entering data values. Once the calculator is in the SD mode, you can accumulate data values by pressing the $\boxed{\text{M+}}$ key after each new value is entered.

Here is a procedure for finding the mean and standard deviation of 1, 2, 5, and 200 that will work for many Casio calculators:

- Press the DATA entry key after entering each of the data values, as in

$$1 \quad \boxed{\text{M+}} \quad 2 \quad \boxed{\text{M+}} \quad 5 \quad \boxed{\text{M+}} \quad 200 \quad \boxed{\text{M+}}.$$

- Find the mean by pressing the $\boxed{\text{SHIFT}}$ key and then pressing the key that has x printed directly above it.
- Find the standard deviation by pressing the $\boxed{\text{SHIFT}}$ key and then pressing the key that has x_{σ_n} printed directly above it.

If you are using a Texas Instruments calculator, you can follow a similar procedure by using the $\boxed{\Sigma+}$ key instead of the $\boxed{M+}$ key. Since not all calculators work in the same way, you may need to read the manual that came with your calculator.

Example

2. On a certain civil service examination, the grades of five people were 71, 73, 74, 86, and 96. Find:
(a) the mean grade
(b) the standard deviation to the *nearest tenth*
(c) the percentage of the total scores that fall within one standard deviation of the mean

Solutions: Using your scientific calculator, enter and accumulate the five data values.
(a) Press the $\boxed{2nd}$ function key, and then press the key that has x printed directly above it. The mean grade is **8.0**.
(b) Press the $\boxed{2nd}$ function key, and then press the key that has σ printed directly above it. The standard deviation, correct to the *nearest tenth*, is **9.6**.
(c) One standard deviation to the left of the mean is $80 - 9.6 = 70.4$, and one standard deviation to the right of the mean is $80 + 9.6 = 89.6$. Four out of five grades are between 70.4 and 89.6. Only 96 is not in this interval.

Hence $\dfrac{4}{5}$ or **80%** of the grades fall within one standard deviation of the mean.

Exercise Set 10.3

1–6. Evaluate.

1. $\displaystyle\sum_{k=1}^{4} \sqrt{2}k^3$

3. $\displaystyle\sum_{k=1}^{5} (k-1)^2$

5. $\displaystyle\sum_{k=1}^{3} \left(\frac{3^k}{k}\right)$

2. $\displaystyle\sum_{k=1}^{5} (k^2 - 1)$

4. $\displaystyle\sum_{n=2}^{4} (n - n^2)$

6. $\displaystyle\sum_{m=1}^{3} (2m+1)^{m-1}$

7–12. Find the numerical value.

7. $\displaystyle\sum_{k=1}^{4} \sin\frac{k\pi}{2}$

8. $\displaystyle\sum_{k=0}^{4} \left(\cos\frac{k\pi}{2}\right)^2$

9. $\displaystyle\sum_{k=0}^{4} \sin\frac{k\pi}{4}$

10. $\displaystyle\sum_{k=1}^{3} \cos\frac{\pi}{k}$ 　　　**11.** $\displaystyle\sum_{p=0}^{3} \log 10^{p}$ 　　　**12.** $\displaystyle\sum_{n=1}^{4} 4^{\frac{n}{2}}$

13. Simplify:

(a) $\displaystyle\sum_{n=0}^{4} i^{n}$ 　and　 (b) $\displaystyle\sum_{n=37}^{40} i^{n}$, 　where $i = \sqrt{-1}$

14. If $\log x = a$, express $\displaystyle\sum_{k=1}^{4} \log x^{k}$ in simplest form in terms of a.

15. Find the value of $\displaystyle\sum_{k=2}^{5} {}_{7}C_{k+1}$.

16. The table below shows the ages at inauguration of ten presidents of the United States.

President	Age at Inauguration (years)
Harry S. Truman	60
Dwight D. Eisenhower	62
John F. Kennedy	43
Lyndon B. Johnson	55
Richard M. Nixon	56
Gerald R. Ford	61
Jimmy Carter	52
Ronald Reagan	69
George H. W. Bush	64
William J. Clinton	46

Find, to the *nearest tenth*, the standard deviation of the ages at inauguration of these ten presidents.

17. A high school football team scored the following numbers of points during the ten-game season: 19, 20, 21, 27, 29, 29, 34, 40, 40, 41.
(a) Find:
(1) the median 　　(2) the mean
(3) the standard deviation to the *nearest hundredth*
(b) What percentage of the total scores fall within one standard deviation of the mean?

10.4 GROUPING DATA

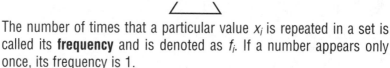

KEY IDEAS

The number of times that a particular value x_i is repeated in a set is called its **frequency** and is denoted as f_i. If a number appears only once, its frequency is 1.

Organizing data values by listing them in a table with their frequencies often simplifies statistical computations.

Interpreting Frequency Tables

Table 10.2 shows members of the set of data values 3.0, 3.0, 4.0, 4.0, 4.0, 5.0, 5.0, 5.0, 5.0, 5.0, organized according to their frequencies. Thus:

$$x_1 = 3.0 \quad \text{and} \quad f_1 = 2$$
$$x_2 = 4.0 \quad \text{and} \quad f_2 = 3$$
$$x_3 = 5.0 \quad \text{and} \quad f_3 = 5$$

TABLE 10.2 DATA VALUES AND FREQUENCIES

x_i Score	f_i Frequency
3.0	2
4.0	3
5.0	5

Since

$$\Sigma f_i = f_1 + f_2 + f_3 = 10,$$

Σf_i represents the total number of data values in the sample.

Examples

1. Using Table 10.2, find:
(a) the median   **(b)** the mean

Solutions: **(a)** Since $\Sigma f_i = 10$, there are 10 scores. The two "middle" scores are the fifth and sixth scores in the table. Since the fifth score is 4.0 and the sixth score is 5.0, the median is $\dfrac{4.0 + 5.0}{2} = \textbf{4.5}$.

(b) Method 1: Divide the sum of the 10 data values by 10:

$$\bar{x} = \frac{\Sigma f_i x_i}{\Sigma f_i} = \frac{2(3.0) + 3(4.0) + 5(5.0)}{2 + 3 + 5} = \frac{43}{10} = \textbf{4.3}$$

Method 2: Using your scientific calculator, enter and accumulate the 10 data values. Press the [2nd] function key, and then press the key that has $\bar{x}$ printed directly above it. The mean is **4.3**.

2. Using the accompanying set of grouped data, find:
(a) the sum of the data values
(b) the mean
(c) the standard deviation correct to the *nearest tenth*
(d) the probability that a measure selected at random will fall within one standard deviation of the mean

x_i Measure	f_i Frequency
50	4
58	4
62	3
64	6
65	2
68	1

Solutions: (a) Using your scientific calculator, enter and accumulate the 20 data values. If you have a Casio calculator, you can quickly enter the same data value more than once by repeatedly pressing the ⌊M+⌋ key. For example, to enter 50 four times, make sure the Casio calculator is in the SD mode and then key in

$$50 \quad \boxed{M+} \quad \boxed{M+} \quad \boxed{M+} \quad \boxed{M+}.$$

You can find the sum of the statistical data values stored in many Casio calculators by pressing the recall ⌊RCL⌋ key followed by the number ⌊2⌋ key. The display shows 1200 and Σx in the corner of the window. Since $\Sigma x = 1200$, the sum of the 20 data values entered is **1200**.

Some scientific calculators have a frequency ⌊FRQ⌋ key that allows you to enter the number of times that a data value occurs in the set. For example, if you have a Texas Instruments calculator with a ⌊FRQ⌋ key, you can enter 50 four times by using this key sequence:

$$50 \quad \boxed{2nd} \quad \boxed{FRQ} \quad 4 \quad \boxed{\Sigma+}.$$

You can find the sum of the statistical data values stored in a Texas Instruments calculator by pressing the ⌊2nd⌋ function key followed by the key that has Σx printed directly above it. If you need help in learning the statistical features of the calculator you are using, check the instruction booklet for that calculator.

(b) To find the mean, press the ⌊2nd⌋ function key and then press the key that has x printed directly above it. The mean is **60**.

(c) To find the standard deviation, press the ⌊2nd⌋ function key and then press the key that has σ printed directly above it. The standard deviation, correct to the *nearest tenth,* is **5.6**.

(d) $P(\text{any } x_i \text{ is within 1 S.D.}) = \dfrac{\text{Number of } x_i \text{ within 1 S.D.}}{\text{Total number of measures}}$

Since the mean is 60 and the standard deviation is approximately 5.6, you need to find the number of measures that are between $60 - 5.6$ and $60 + 5.6$ or, equivalently, between 54.4 and 65.6. The data values and their corresponding frequencies that fall in this interval are as follows:

$$x_2 = 58, f_2 = 4; \quad x_3 = 62, f_3 = 3; \quad x_4 = 64, f_4 = 6; \quad x_5 = 65, f_5 = 2.$$

The total number of data values that fall within one standard deviation of the mean is

$$f_2 + f_3 + f_4 + f_5 = 4 + 3 + 6 + 2 = 15.$$

Since the total number of measures is 20,

$$P(x_i \text{ is within } 1 \text{ S.D.}) = \frac{15}{20} = \frac{3}{4}.$$

Exercise Set 10.4

1–3. For each set of measurements, find:

(a) the median
(b) the mean to the *nearest tenth*

(c) the standard deviation to the *nearest tenth*
(d) the probability that a measure picked at random will fall within one standard deviation of the mean

1.

x_i	f_i
60	1
75	7
80	3
90	4

2.

x_i	f_i
91	2
96	3
105	6
111	3
113	4

3.

x_i	f_i
56	4
58	3
60	1
61	5
62	7

4. The accompanying table shows the grades of 20 students on a Course III math exam.
   (a) Find, to the *nearest tenth*, the standard deviation.
   (b) Which statement is true for this set of data?
      (1) median > mode
      (2) median = mode
      (3) median < mode
   (c) What is the probability that a score selected at random lies within one standard deviation of the mean?

Score	Frequency
72	6
76	5
84	7
88	2

10.5 WORKING WITH NORMAL CURVES

If each of the different heights of the students in your school is plotted against the corresponding number of students at each height, a curve that is approximately bell-shaped will most likely result. Many real-life data, when graphed, follow a bell-shaped curve.

If you know that a set of data "fits" a bell-shaped, or *normal*, curve, you can draw conclusions about how far the data are from the mean.

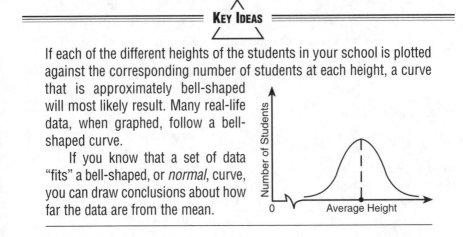

Normal Curves

A bell-shaped curve is actually the graph of a complicated mathematical function that is studied in more advanced statistics courses. The graph of this function is called a **normal curve**. A normal curve is symmetric with respect to the vertical line $x = x$. If a set of data values, such as the scores on a test, fits a normal curve, then the numbers of these data values that fall within one, two, and three standard deviations of the mean can be predicted, as shown in the accompanying figure.

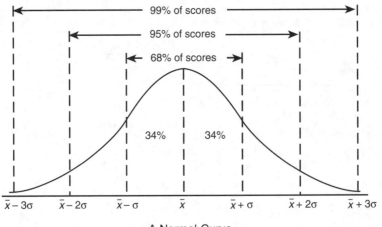

A Normal Curve

Since 68% of the data values or scores fall within one standard deviation of the mean, 34% fall on either side of (one standard deviation away from) the mean. Since the normal curve is symmetric about the vertical line $x = x$, the mean and the median coincide. Hence 50% of the data values are found on either side of the mean.

When a large number of plotted data points closely approximate the bell shape of a normal curve, the data are said to be *approximately* **normally distributed**. When data are *assumed* to be normally distributed, the relationships summarized in Table 10.3 can be used to draw conclusions concerning the approximate numbers of data scores that are one, two, and three standard deviations from the mean.

TABLE 10.3 NORMALLY DISTRIBUTED SCORES [σ REPRESENTS THE STANDARD DEVIATION (S.D.)]

Interval	Interval Length	Contains . . .
$x - \sigma$ to $x + \sigma$	2σ	68% of all scores
$x - 2\sigma$ to $x + 2\sigma$	4σ	95% of all scores
$x - 3\sigma$ to $x + 3\sigma$	6σ	99% of all scores

Examples

1. A set of test scores that follows a bell curve has a mean of 65 and a standard deviation of 5.
(a) Approximately what percent of the scores fall between 55 and 75?
(b) If 1000 students took the test, about how many students scored between 60 and 70?

Solutions: **(a)** If one standard deviation is 5, then an interval of two standard deviations is 10. Since $65 - 10 = 55$ and $65 + 10 = 75$, the interval 55 to 75 represents *two* standard deviations on either side of the mean of 65.

Since the data follow a bell curve, you can expect approximately **95%** of the test scores to fall between 55 and 75.

(b) Since $65 - 5 = 60$ and $65 + 5 = 70$, the interval 60 to 70 represents one standard deviation about the mean of 65. Hence about 68% of the test scores fall between 60 and 70.

If 1000 students took the test, then approximately 68% × 1000 or **680** students had scores between 60 and 70.

2. On a quiz, the mean score is 25 and the standard deviation is 2.3. Which score could be expected to occur less than 5% of the time?
(1) 20   (2) 28   (3) 23   (4) 24

Solution: If it is assumed that the scores are normally distributed, 95% of them are expected to fall within two standard deviations of the mean, so scores that are *more than* two standard deviations from the mean are expected to occur less than 5% of the time. Since an interval of two standard deviations is $2 \times 2.3 = 4.6$, look in the choices for a score that differs from the mean of 25 by more than 4.6. The score of 20, choice (1), is the only value given that is more than 4.6 units from 25.

The correct choice is **(1)**.

3. On an exam, 95% of the scores ranged between 74 and 84. If the scores are approximately normally distributed and the mean is 79, find the standard deviation.

Solution: The length of the interval from 74 and 84 is 10 points. Since 95% of normally distributed scores fall within two standard deviations of the mean, the length of the interval from 74 to 79 is 2σ and from 79 to 84 is another 2σ. Thus, the length of the interval 74 to 84 is a total of 4σ.

$$4\sigma = 10$$
$$\sigma = \frac{10}{4} = \textbf{2.5}$$

4. The scores on a standardized exam have a mean of 78 and a standard deviation of 3.5. If a normal distribution of scores is assumed, a student's score of 86 would rank:
(1) below the 75th percentile
(2) between the 75th and 85th percentiles
(3) above the 95th percentile
(4) between the 95th and 85th percentiles

Solution: Since two standard deviations represent $2 \times 3.5 = 7$ points, a score of 85 is two standard deviations to the *right* of the mean of 78. Although 95% of the scores fall within two standard deviations of *either side* of the mean, half of this number (47.5%) falls to the *right* of the mean.

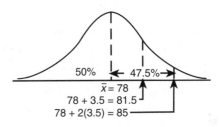

In a normal distribution 50% of the scores fall to the *left* of the mean. This means that a total of approximately 50% + 47.7%, or 97.5%, of the scores will be less than or equal to 85. A score of 86 is, therefore, above the 95th percentile.

The correct choice is **(3)**.

Exercise Set 10.5

1. Approximately what percent of normally distributed scores fall between 70 and 90 if the mean is 80 and the standard deviation is:
(a) 10?   (b) 5?   (c) 3.3?

2. A set of standardized test scores has a mean of 73 and a standard deviation of 7.
   (a) Approximately what percent of the scores fall between 66 and 80?
   (b) If 1000 students took the test, about how many students scored between:
      (1) 66 and 80?   *(2)* 73 and 80?   *(3)* 59 and 87?

3. On a quiz, the mean score is 72 and the standard deviation is 3.4. Which score could be expected to occur *less than* 5% of the time?
   (1) 65   (2) 67   (3) 72   (4) 78

4. The scores on an exam are normally distributed with a mean of 77. Find the standard deviation if the interval from 71 to 83 contains:
   (a) 68% of the scores   **(b)** 95% of the scores   **(c)** 99% of the scores

5. Which of the following statements about a normal distribution is *always* true?
   (1) the mean = the median
   (2) the mean < median
   (3) the mean > the median
   (4) the variance > the standard deviation

6. In the accompanying diagram, the shaded area represents approximately 95% of the scores on a standard test. If these scores ranged from 52 to 96, which of the following could be the standard deviation?
   (1) 11   (3) 76
   (2) 22   (4) 44

7. A student scores 84 on a standardized exam. The mean for the exam is 75, and the standard deviation is 4. If a normal distribution of scores is assumed, how does this student rank?
   (1) below the 75th percentile
   (2) between the 75th and 85th percentiles
   (3) above the 95th percentile
   (4) between the 85th and 95th percentiles

8. The heights of the members of a high school class are normally distributed. If the mean height is 65 inches and a height of 72 inches represents the 84th percentile, what is the standard deviation for this distribution?
   (1) 7   (2) 14   (3) 12   (4) 137

9. The mean for a standardized exam is 81, and the standard deviation is 6. If the scores are normally distributed, approximately what percentile corresponds to each of the following scores?
   (a) 81   (b) 75   (c) 93   (d) 87

10. For a standardized test whose scores are normally distributed, if the mean is 20 and the standard deviation is 2.6, which score will occur less than 5% of the time?
    (1) 25.1   (2) 22.6   (3) 18.7   (4) 14.6

REGENTS TUNE-UP: CHAPTER 10

Each of the questions in this section has appeared on a previous Course III Regents Examination. Here is an opportunity for you to review Chapter 10 and, at the same time, prepare for the Course III Regents Examination.

1. Evaluate $\displaystyle\sum_{k=2}^{4}(4-k^2)$.

2. On a standardized test, the mean is 83 and the standard deviation is 3.5. What is the best approximation of the percentage of scores that fall in the range 76–90?
   (1) 34   (2) 68   (3) 95   (4) 99

3. The mean of a normally distributed set of data is 52, and the standard deviation is 4. Approximately 95% of all cases will lie between which measures?
   (1) 44 and 52   (2) 44 and 60   (3) 48 and 56   (4) 52 and 64

4. What is the last term of the expansion of $(2x - 3y)^4$?

5. A fair coin is tossed ten times, and a head appears each time. What is the probability that on the next three tosses of the coin *exactly* two heads will appear?

6. If the probability of winning a game is $\dfrac{1}{4}$, find the probability of winning at least three games out of four.

7. What is the seventh term in the expansion of $(2x - y)^7$?

8. (a) Find, to the *nearest tenth*, the standard deviation for the accompanying data.
   (b) The given data do *not* form a normal distribution. If a measure is selected at random from these data, what is the probability that it will differ from the mean by less than one standard deviation?

Measure (x_i)	Frequency (f_i)
20	4
21	2
24	5
26	3
30	6

9. The table below shows the heights of a group of 20 students.

(a) Find the mean and the standard deviation to the *nearest tenth*.

(b) If one student's height is chosen at random, what is the probability that the height falls within one standard deviation of the mean?

(c) If three students' heights are chosen at random, what is the probability that *at most* one of them falls within one standard deviation of the mean?

Height (inches)	Frequency
72	3
71	2
70	1
69	2
68	4
67	2
66	4
65	2

10. In the accompanying diagram, the circle is divided into four sections as shown, and

$$m\overset{\frown}{AB} : m\overset{\frown}{BC} : m\overset{\frown}{CD} : m\overset{\frown}{DA} = 3 : 4 : 2 : 1$$

(a) If the spinner is spun once, find:
(1) $P(\text{red})$ *(2)* $P(\text{green})$

(b) Determine the probability of obtaining:
(1) exactly two greens in three spins
(2) at least three reds in four spins
(3) at most two yellows in three spins

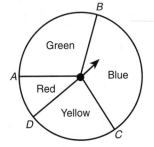

ANSWERS TO SELECTED EXERCISES: CHAPTER 10

Section 10.1

1. $\dfrac{54}{125}$

2. (3)

3. (2)

4. (3)

5. (3)

6. (1)

7. .3

8. $\dfrac{51}{243}$

9. (a) $\dfrac{343}{13,824}$ (b) $\dfrac{20}{27}$

10. $\dfrac{41}{59,049}$

11. (a) .2401   (b) .3773

Section 10.2

1. $x^4 + 4x^3y + 6x^2y^2 + 4xy^3 + y^4$

2. $p^4 - 12p^3q + 54p^2q^2 - 108pq^3 + 81q^4$

3. (3)
4. (1)
5. (1)
6. (4)
7. $28p^6q^2$
8. $30,375x^4y^2$
9. $-280x^4$
10. $-\dfrac{20}{8}a^3b^3$

11. $80rs^4$
12. $35a^3b^4$
13. 15
14. (a) -32
    (b) $\dfrac{6}{243}bc^5$

15. (a) $-150x^2$
    (b) $-5x^3$
16. (a) $-252a^5b^5i$
    (b) $-19ab^9i$

Section 10.3

1. $100\sqrt{2}$
2. 50
3. 30
4. -20
5. $\dfrac{33}{2}$
6. 55

7. 0
8. 3
9. $1+\sqrt{2}$
10. $-\dfrac{1}{2}$
11. 6
12. 30

13. (a) 1
    (b) 0
14. 10a
15. 98
16. 7.7

17. (a) *(1)* 29
        (2) 30
        (3) 8.06
    (b) 40%

Section 10.4

1. (a) 75
    (b) 79
    (c) 8
    (d) $\dfrac{10}{15}$

2. (a) 105
    (b) 104.7
    (c) 7.4
    (d) $\dfrac{9}{18}$

3. (a) 61
    (b) 59.9
    (c) 2.3
    (d) $\dfrac{16}{20}$

4. (a) 5.8
    (b) *(3)*
    (c) $\dfrac{12}{20}$

Section 10.5

1. (a) 68%   (b) 95%   (c) 99%
2. (a) 68%
    (b) *(1)* 680   *(2)* 340   *(3)* 950
3. (1)
4. (a) 6   (b) 3   (c) 2
5. (1)   **6.** (1)   **7.** (3)   **8.** (1)
9. (a) 50   (b) 16   (c) 95   (d) 84
10. (4)

Regents Tune-Up: Chapter 10

1. -17
2. (3)
3. (2)
4. $81y^4$

5. $\dfrac{3}{8}$

6. $\dfrac{13}{256}$

7. $14xy^6$

8. (a) 3.8

   (b) $\dfrac{8}{20}$

9. (a) $\bar{x} = 68.3,\ \sigma = 2.3$

   (b) $\dfrac{13}{20}$

   (c) $\dfrac{2254}{8000}$

10. (a) *(1)* $\dfrac{1}{10}$ *(2)* $\dfrac{3}{10}$

   (b) *(1)* $\dfrac{189}{1000}$ *(3)* $\dfrac{124}{125}$

   (2) $\dfrac{37}{10000}$

GLOSSARY

Abscissa The x-coordinate of a point in the coordinate plane.

Absolute value The absolute value of a number x, denoted by $|x|$, is its distance from zero on the number line. Thus, $|x|$ always represents a nonnegative number.

Acute angle An angle whose degree measure is greater than 0 but less than 90.

Acute triangle A triangle that contains three acute angles.

Additive inverse The opposite of a number. The additive inverse of a number x is $-x$ since $x + (-x) = 0$.

Altitude A segment that is perpendicular to the side to which it is drawn.

Ambiguous case The case in which the measures of two sides and an angle that is not included between the two sides are given. These measures may determine one triangle, two triangles, or no triangle.

Amplitude The amplitude of functions of the form $y = a \sin bx$ and $y = a \cos bx$ refers to their maximum height, which is $|a|$.

Angle The union of two rays that have the same endpoint.

Angle of depression The angle formed by a horizontal line of vision and the line of sight by which an object is viewed beneath the horizontal line of vision.

Angle of elevation The angle formed by a horizontal line of vision and the line of sight by which an object is viewed above the horizontal line of vision.

Antilogarithm The antilogarithm of a given number x is the number N whose logarithm is x. Thus, the antilogarithm of 2 is 100 since $\log 100 = 2$.

Arc A part of a circle. If the degree measure of the arc is less than 180, the arc is a **minor** arc. If the degree measure of the arc is greater than 180, the arc is a **major** arc. A **semicircle** is an arc whose degree measure is 180.

Arc cos x or Cos⁻¹ x The angle A whose cosine is x, where $0° \le A \le 180°$.

arc cos x or cos⁻¹ x An angle whose cosine is x, where the angle is not restricted in value.

Arc sin x or Sin⁻¹ x The angle A whose sine is x, where $-90° \le A \le 90°$.

arc sin x or sin⁻¹ x An angle whose sine is x, where the angle is not restricted in value.

Arc tan x or Tan⁻¹ x The angle A whose tangent is x, where $-90° < A < 90°$.

arc tan x or tan⁻¹ x An angle whose tangent is x, where the angle is not restricted in value.

Area of a triangle The number of square units enclosed by the triangle, which can be computed by one of these two formulas:

(1) Area = $\frac{1}{2} bh$, where b is a side of the triangle, called the base, and h is the altitude drawn to that side;

(2) Area = $\frac{1}{2} ab \sin C$, where a and b are the lengths of any two sides of the triangle and C is the angle of the triangle formed by those two sides.

Associative property The mathematical law that states that the way in which three numbers are grouped when they are added or multiplied does not matter.

Asymptote A line is a horizontal asymptote if a graph approaches it but does not intersect it as x increases or decreases without bound.

Average See **Mean**.

Bernoulli experiment The repetition of a simple probability experiment having exactly two possible outcomes, for a fixed number of times under identical conditions, so that the outcome of any one experiment, called a *trial,* does not affect the outcome of any other trial.

Binomial A polynomial with two unlike terms, as in $3x + 2y$.

Binomial theorem A formula that tells how to expand a binomial of the form $(a + b)^n$ without performing repeated multiplications, as in
$$(a + b)^3 = {}_3C_0a^3$$
$$+ {}_3C_1a^{3-1} b^1 + {}_3C_2a^{3-2} b^2 + {}_3C_3a^{3-3}b^3$$
$$= a^3 + 3a^2b + 3ab^2 + b^3$$

Central angle An angle whose vertex is the center of a circle and whose sides are radii.

Characteristic The integer part of a common logarithm.

Chord A line segment whose endpoints are points on a circle.

Circle The set of points (x, y) in a plane that are a fixed distance r from a given point (h, k), called the *center*. Thus, $(x - h)^2 + (y - k)^2 = r^2$.

Combination A subset of a set of objects in which order is not considered.

Combination formula The combination of n objects taken r at a time, denoted by ${}_nC_r$, is given by the formula ${}_nC_r = \frac{{}_nP_r}{r!} = \frac{n!}{r!(n-r)!}$

Common logarithm A logarithm whose base is 10.

Commutative property The mathematical law that states that the order in which two numbers are added or multiplied does not matter.

Complementary angles Two angles whose degree measures add up to 90.

Complex fraction A fraction that contains other fractions in its numerator, its denominator, or both its numerator and its denominator.

Composition of functions The composition of function f followed by function g is a new function, denoted by g ∘ f, consisting of the set of function values g(f(x)), provided that f(x) is in the domain of g.

Composition of transformations A sequence of transformations in which the image of one transformation is used as the preimage for a second transformation.

Congruent angles (or sides) Angles (sides) that have the same measure. The symbol for congruence is ≅.

Conjugate pair The sum and difference of the same two terms, as in $a + b$ and $a - b$.

Constant A quantity that is fixed in value. In the equation $y = x + 3$, x and y are variables and 3 is a constant.

Coordinate The real number that corresponds to the position of a point on a number line.

Coordinate plane The region formed by a horizontal number line and a vertical number line intersecting at their zero points.

Cosecant The reciprocal of the sine function.

Cosine ratio In a right triangle, the ratio of the length of the leg that is adjacent to a given acute angle to the length of the hypotenuse. If an angle is in standard position, the cosine ratio is $\frac{x}{r}$, where $P(x, y)$ is an arbitrary point on the terminal ray of the angle and $r = \sqrt{x^2 + y^2}$.

Cotangent The reciprocal of the tangent function.

Coterminal angles Angles in standard position whose terminal rays coincide.

Degree A unit of angle measure that is defined as 1/360 of one complete rotation of a ray about its vertex. For conversion of degrees to radians, and of radians to degrees, see **Radian.**

Degree of a monomial The sum of the exponents of the variable factors.

Degree of a polynomial The greatest degree of the monomial terms.

Dependent variable For a function of the form $y = f(x)$, y is the dependent variable.

Diameter A chord of a circle that contains the center of the circle.

Dilation A transformation in which a figure is enlarged or reduced based on a center and a scale factor.

Direct isometry An isometry that preserves orientation.

Discriminant The quantity $b^2 - 4ac$ that is underneath the radical sign in the quadratic formula.

Distributive property of multiplication over addition For any real numbers a, b, and c, $a(b + c) = ab + ac$ and $(b + c) a = ba + ca$.

Domain of a relation The set of all possible first members of the ordered pairs that comprise a relation.

Domain of a variable The set of all possible replacements for a variable.

Equation A statement that two quantities have the same value.

Equivalent equations Two equations that have the same solution set. Thus, $2x = 6$ and $x = 3$ are equivalent equations.

Ellipse An oval curve an equation of which is $ax^2 + by^2 = c$, where a, b, and c have the same sign and $a \neq b$.

Event A particular subset of outcomes from the set of all possible outcomes of a probability experiment. In flipping a coin, one event is getting a head; another event is getting a tail.

Exponent In x^n, the number n is the exponent and tells the number of times the base x is used as a factor in a product. Thus, $x^3 = (x)(x)(x)$.

Exponential equation An equation in which the variable appears in an exponent.

Exponential function A function of the form $y = b^x$, where b is a positive constant that is not equal to 1.

Extremes In the proportion $\frac{a}{b} = \frac{c}{d}$, the terms a and d are the extremes.

Factor A number or variable that is being multiplied in a product.

Factorial n For any positive integer n, factorial n is denoted by $n!$ and is defined as the product of consecutive integers from n to 1. Thus, $5! = 5 \cdot 4 \cdot 3 \cdot 2 \cdot 1 = 120$.

Factoring The process by which a number or polynomial is written as the product of two or more terms.

Factoring completely Factoring a number or polynomial into its prime factors.

Failure In a two-outcome probability experiment, the outcome that is not a "success" is a "failure."

FOIL The rule for multiplying two binomials horizontally by forming the sum of the products of the first terms (F), the outer terms (O), the inner terms (I), and the last terms (L) of each binomial.

Frequency The number of times a data value occurs or the number of full cycles the graph of a trigonometric function completes in an interval of 2π radians.

Function A relation in which no two ordered pairs have the same first member but different second members.

Fundamental counting principle If event A can occur in m ways and event B can occur in n ways, then both events can occur in m times n ways.

Glide reflection The composite of a line reflection and a translation whose direction is parallel to the reflecting line.

Greatest common factor (GCF) The GCF of two or more monomials is the monomial with the largest coefficient and the variable factors of the greatest degree that are common to all the given monomials. Thus, the GCF of $8a^2b$ and $20ab^2$ is $4ab$.

Half-turn A rotation in which the angle of rotation is $180°$.

Horizontal-line test If no horizontal line intersects the graph of a function in more than one point, the graph represents a one-to-one function.

Hyperbola A curve that consists of two branches, an equation of which is $ax^2 + by^2 = c$, where a and b have opposite signs and $c \neq 0$. A special type of hyperbola, called an *equilateral* or *rectangular hyperbola*, has the equation $xy = k$, where $k \neq 0$.

Hypotenuse The side of a right triangle that is opposite the right angle.

Image In a geometric transformation, the point or figure that corresponds to the original point or figure.

Imaginary number A number of the form $a + bi$, where b is a nonzero real number and i is the imaginary unit.

Imaginary unit The number denoted by i, where $i = \sqrt{-1}$.

Independent variable For a function of the form $y = f(x)$, x is the independent variable.

Index The number k in the radical expression $\sqrt[k]{x}$, which tells what root of x is to be taken. In a square root radical the index is omitted and is understood to be 2.

Inequality A sentence that uses an inequality relation such as < (is less than), ≤ (is less than or equal to), > (is greater than), ≥ (is greater than or equal to), or ≠ (is unequal to).

Initial side The side of an angle in standard position that remains fixed on the positive x-axis.

Inscribed angle An angle whose vertex is a point on a circle and whose sides are chords.

Integer A number from the set $\{\ldots, -3, -2, -1, 0, 1, 2, 3, \ldots\}$.

Inverse variation A set of ordered pairs in which the product of the first and second members of each pair is the same non-zero number. Thus, if y varies indirectly as x, then $xy = k$, where k is a nonzero number called the constant of variation.

Irrational number A number that cannot be expressed as the quotient of two integers.

Isometry A transformation that preserves distance.

Isosceles triangle A triangle in which two sides have the same length. The unequal side is called the **base** and the equal sides are called **legs.** The angle formed by the legs is called the **vertex angle.**

Law of Cosines In a triangle, the square of the length of any side is equal to the sum of the squares of the lengths of the other two sides minus twice the product of the lengths of these sides and the cosine of the included angle.

Law of Sines In a triangle the ratio of the length of any side to the sine of the angle opposite that side is the same.

Leg of a right triangle Either of the two sides of a right triangle that is not opposite the right angle.

Linear equation An equation in which the greatest exponent of a variable is 1. A linear equation in two variables can be put into the form $Ax + By = C$, where A, B, and C are constants and A and B are not both zero.

Line reflection A transformation in which each point P that is not on line ℓ is paired with a point P' on the opposite side of line ℓ so that line ℓ is the perpendicular bisector of $\overline{PP'}$. If P is on line ℓ, then P is paired with itself.

Line symmetry A figure has line symmetry when a line ℓ divides the figure into two parts such that each part is the reflection of the other part in line ℓ.

Logarithm of x An exponent that represents the power to which a given base must be raised to produce a positive number x. Thus, $\log_2 8$ is 3 since $2^3 = 8$.

Major arc An arc whose degree measure is greater than 180.

Mapping A relation in which each member of one set is paired with exactly one member of a second set.

Mean (Average) The value obtained when the sum of a set of n data values is divided by n.

Means In the proportion $\dfrac{a}{b} = \dfrac{c}{d}$, the terms b and c are the means.

Median The middle value when the numbers in a set of data values are arranged in size order.

Median of a triangle A line segment whose endpoints are a vertex of the triangle and the midpoint of the opposite side.

Minor arc An arc whose degree measure is less than 180.

Mode The value that occurs most frequently in a given set of data.

Monomial A number, a variable, or the product of a number and a variable, as in $3x^2y$.

Multiplicative inverse The reciprocal of a nonzero number.

Negative angle An angle in standard position whose terminal ray rotates in a clockwise direction.

Normal curve A bell-shaped curve that describes a distribution of data values in which approximately 68 percent of the values fall within one standard deviation of the mean, 95 percent fall within two standard deviations of the mean, and 99 percent fall within three standard deviations of the mean.

Obtuse angle An angle whose degree measure is greater than 90 but less than 180.

Obtuse triangle A triangle that contains an obtuse angle.

One-to-one function A function in which no two ordered pairs have the same value of y but different values of x.

Ordered pair Two numbers that are written in a definite order.

Ordinate The y-coordinate of a point in the coordinate place.

Origin The zero point on a number line.

Outcome A possible result in a probability experiment.

Perfect square A rational number whose square root is rational, as in 25 or $\frac{9}{16}$.

Period The period of a sine, cosine, or tangent curve is the number of degrees or radians needed for the curve to complete one full cycle. The period of functions of the form

$y = a \sin bx$ and $y = a \cos bx$ is $\frac{2\pi}{|b|}$.

Permutation An ordered arrangement of objects.

Point symmetry A figure has point symmetry if, after it has been rotated 180°, the image coincides with the original figure.

Polygon A simple closed curve whose sides are line segments.

Polynomial A monomial or the sum or difference of two or more monomials, as in $3x + 2y$.

Positive angle An angle in standard position whose terminal ray rotates in a counter-clockwise direction.

Power A number written with an exponent.

Preimage If under a certain transformation A' is the image of A, then A is the preimage of A'.

Prime factorization The factorization of a polynomial into factors each of which is divisible only by itself and 1 (or -1).

Probability of an event The number of ways in which the event can occur divided by the total number of possible outcomes.

Proportion An equation that states that two ratios are equal. In the proportion $\frac{a}{b} = \frac{c}{d}$, the product of the means equals the product of the extremes: $b \cdot c = a \cdot d$.

Pythagorean theorem The square of the length of the hypotenuse of a right triangle is equal to the sum of the squares of the lengths of the legs of the triangle.

Quadrant One of four rectangular regions into which the coordinate plane is divided.

Quadrantal angle An angle in standard position whose terminal ray coincides with a coordinate axis.

Quadratic equation An equation that can be put into the form $ax^2 + bx + c = 0$, provided that $a \neq 0$.

Quadratic formula The roots of the quadratic equation $ax^2 + bx + c = 0$ are given by the formula $x = \frac{-b \pm \sqrt{b^2 - 4ac}}{2a}$ $(a \neq 0)$.

Quadratic polynomial A polynomial whose degree is 2.

Quadrilateral A polygon with four sides.

Radian The measure of a central angle of a circle that intercepts an arc whose length equals the radius of the circle. To change from degrees to radians, multiply the number of degrees by $\frac{\pi}{180°}$. To change from radians to degrees, multiply the number of radians by $\frac{180°}{\pi}$.

Radical equation An equation in which the variable appears underneath the radical sign.

Radical (square root) sign The symbol $\sqrt{\ }$, which denotes the positive square root of a nonnegative number.

Radicand The expression that appears underneath a radical sign.

Range (1) The set of all possible second members of the ordered pairs that comprise a relation. (2) In a set of numerical data values, the difference between the greatest and smallest values.

Ratio A comparison of two numbers by division. The ratio of a to b is the fraction $\frac{a}{b}$, provided that $b \neq 0$.

Rational number A number that can be written in the form $\frac{a}{b}$, where a and b are integers and $b \neq 0$. Decimals in which a set of digits endlessly repeat, such as $0.25000\ldots \left(=\frac{1}{4}\right)$ and $0.33333\ldots \left(=\frac{1}{3}\right)$, represent rational numbers.

Real number A number that is a member of the set that consists of all rational and irrational numbers.

Reference angle When an angle is placed in standard position, the acute angle formed by the terminal ray and the x-axis.

Relation A set of ordered pairs.

Replacement set The set of values that a variable may have.

Right angle An angle whose degree measure is 90.

Right triangle A triangle that contains a right angle.

Root A number that makes an equation a true statement.

Rotation A transformation in which a point or figure is moved about a fixed point a given number of degrees.

Rotational symmetry A figure has rotational symmetry if, after a rotation of more than 0° but less than 360°, the image coincides with the original figure.

Scalene triangle A triangle in which the three sides have different lengths.

Secant The reciprocal of the cosine function.

Secant line A line that intersects a circle in two different points.

Semicircle An arc whose degree measure is 180.

Sigma (lower case) The Greek letter σ, which represents standard deviation.

Sigma (upper case) The Greek letter Σ, which indicates a summation of terms.

Similar figures Figures that have the same shape. Two triangles are similar if two pairs of corresponding angles have the same degree measure.

Sine ratio In a right triangle, the ratio of the length of the leg that is opposite a given acute angle to the length of the hypotenuse. If an angle is in standard position, the sine ratio is $\frac{y}{r}$, where $P(x, y)$ is an arbitrary point on the terminal ray of the angle and $r = \sqrt{x^2 + y^2}$.

Solution Any value from the replacement set of a variable that makes an open sentence true.

Solution set The collection of all values from the replacement set of a variable that make an open sentence true.

Square root The square root of a nonnegative number n is one of two identical numbers whose product is n. Thus, 3 and -3 are the two square roots of 9 since $3 \times 3 = 9$ and $(-3) \times (-3) = 9$. The symbol $\sqrt{9}$ represents the principal or positive square root of 9, so $\sqrt{9} = 3$ because $3 \times 3 = 9$.

Standard deviation The square root of the sum of the squares of the differences between all of the individual data values and the mean, divided by the number of data values.

Standard position An angle whose vertex is fixed at the origin and whose initial side coincides with the positive x-axis.

Success Any favorable outcome of a probability experiment.

Supplementary angles Two angles whose degree measures add up to 180.

System of equations A set of equations whose solution is the set of values that make all of the equations true at the same time.

Tangent line A line that intersects a circle in exactly one point.

Tangent ratio In a right triangle, the ratio of the length of the leg that is opposite a given acute angle to the length of the leg that is adjacent to that angle. If an angle is in standard position, the tangent ratio is $\frac{y}{x}(x \neq 0)$, where $P(x, y)$ is an arbitrary point on the terminal ray of the angle.

Terminal ray The side of an angle in standard position that rotates about the origin.

Theorem A mathematical generalization that can be proved.

Transformation A one-to-one mapping whose domain and range are the set of all points in the plane.

Translation A transformation in which each point of a figure is moved the same distance and in the same direction.

Trinomial A polynomial with three unlike terms, as in $x^2 - 3x + 5$.

Unit circle A circle whose radius is 1.

Vertical-line test If no vertical line intersects a graph in more than one point, the graph represents a function.

REGENTS EXAMINATION FORMULAS

To the Student Reader: *The following set of formulas is provided in the New York State Regents Examination question booklet so that you do not have to memorize these formulas for this examination. Your teacher, however, may require you to memorize the formulas for class tests.*

Formulas

Pythagorean and Quotient Identities

$$\sin^2 A + \cos^2 A = 1$$

$$\tan^2 A + 1 = \sec^2 A$$

$$\cot^2 A + 1 = \csc^2 A$$

$$\tan A = \frac{\sin A}{\cos A}$$

$$\cot A = \frac{\cos A}{\sin A}$$

Functions of the Sum of Two Angles

$$\sin(A + B) = \sin A \cos B + \cos A \sin B$$
$$\cos(A + B) = \cos A \cos B - \sin A \sin B$$

$$\tan(A + B) = \frac{\tan A + \tan B}{1 - \tan A \tan B}$$

Functions of the Difference of Two Angles

$$\sin(A - B) = \sin A \cos B - \cos A \sin B$$
$$\cos(A - B) = \cos A \cos B + \sin A \sin B$$

$$\tan(A - B) = \frac{\tan A - \tan B}{1 + \tan A \tan B}$$

Law of Sines

$$\frac{a}{\sin A} = \frac{b}{\sin B} = \frac{c}{\sin C}$$

Law of Cosines

$$a^2 = b^2 + c^2 - 2bc \cos A$$

Functions of the Double Angle

$$\sin 2A = 2 \sin A \cos A$$
$$\cos 2A = \cos^2 A - \sin^2 A$$
$$\cos 2A = 2 \cos^2 A - 1$$
$$\cos 2A = 1 - 2 \sin^2 A$$

$$\tan 2A = \frac{2 \tan A}{1 - \tan^2 A}$$

Functions of the Half Angle

$$\sin \frac{1}{2} A = \pm \sqrt{\frac{1 - \cos A}{2}}$$

$$\cos \frac{1}{2} A = \pm \sqrt{\frac{1 + \cos A}{2}}$$

$$\tan \frac{1}{2} A = \pm \sqrt{\frac{1 - \cos A}{1 + \cos A}}$$

Area of Triangle

$$K = \frac{1}{2} ab \sin C$$

Standard Deviation

$$S.D. = \sqrt{\frac{1}{n} \sum_{i=1}^{n} (x_i - \bar{x})^2}$$

Examination
June 2001
Sequential Math Course III

PART I

Answer 30 questions from this part. Each correct answer will receive 2 credits. No partial credit will be allowed. Write your answers in the spaces provided. Where applicable, answers may be left in terms of π or in radical form. [60]

1 Express $300°$ in radian measure. 1 _____

2 If $f(x) = \sqrt{29 - x^2}$, find f(–2). 2 _____

3 In $\triangle ABC$, $\sin A = \dfrac{1}{4}$, $\sin B = \dfrac{1}{8}$, and $b = 20$. What is the length of a? 3 _____

4 If the number 0.00416 is expressed in scientific natotion as 4.16×10^x, what is the value of x? 4 _____

5 If $f(x) = 2x + 4$ and $g(x) = x^2 + 1$, find $(f \circ g)(3)$. 5 _____

6 In which quadrant will the image of $A(4,-2)$ lie after dilation D_{-2}? 6 _____

7 Solve for x: $2^{4x-1} = 4^x$ 7 _____

8 In $\triangle ABC$, $m\angle C = 30$ and $a = 24$. If the area of the triangle is 42, what is the length of side b? 8 _____

9 What is the value of x in the equation $3^x = 148$, expressed to the *nearest hundredth*?

9 _____

10 Solve for all values of x: $|3x + 5| = 7$

10 _____

11 In the accompanying diagram of circle O, chords $\overline{AB}$ and $\overline{CD}$ intersect at E, $m\widehat{AC} = 50$, and $m\widehat{BD} = 150$. Find $m\angle AED$.

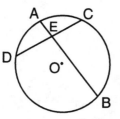

11 _____

12 In $\triangle ABC$, $a = 6$, $b = 10$, and $m\angle C = 120$. What is the length of c?

12 _____

13 The probability of Rick getting an A on any test is $\dfrac{2}{3}$. Find the probability that he earns an A on *exactly* 3 of 4 tests.

13 _____

14 Evaluate: $\displaystyle\sum_{k=0}^{3}(3k - 2)^2$

14 _____

15 In the accompanying diagram, tangent $\overline{AB}$ and secant $\overline{ACD}$ are drawn to a circle. If $AC = 4$ and $CD = 12$, find AB.

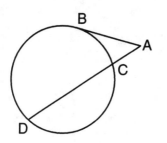

15 _____

Directions (16–35): For *each* question chosen, write in the space provided the *numeral* preceding the word or expression that best completes the statement or answers the question.

16 What is the image of point (2,4) under the translation $T_{-6,1}$?

(1) (−4,3)   (3) (8,3)

(2) (−4,5)   (4) (8,5)   16 _____

17 The expression $3\sqrt{-18}+5\sqrt{-12}$ is equivalent to

(1) $9i\sqrt{2}+10i\sqrt{3}$ (3) $19i\sqrt{5}$

(2) $6i\sqrt{2}+7i\sqrt{3}$ (4) $-90\sqrt{6}$ 17 _____

18 The fraction $\dfrac{3-x}{2x-6}$, $x \neq 3$, is equivalent to

(1) $\dfrac{1}{2}$ (3) $\dfrac{1}{4}$

(2) $-\dfrac{1}{2}$ (4) $-\dfrac{1}{4}$ 18 _____

19 If x is a real number, what is the solution set of the equation $\sqrt{1-2x} = 2$?

(1) $\left\{\dfrac{3}{2}\right\}$ (3) $\{-2\}$

(2) $\left\{-\dfrac{3}{2}\right\}$ (4) $\{\ \}$ 19 _____

20 The value of $\sin \dfrac{4\pi}{3}$ is

(1) $\dfrac{1}{2}$ (3) $\dfrac{\sqrt{3}}{2}$

(2) $-\dfrac{1}{2}$ (4) $-\dfrac{\sqrt{3}}{2}$ 20 _____

21 The expression $\sin^2 x + \cos^2 x - b^2$ is equivalent to

(1) 1   (3) $(1 + b)(1 - b)$
(2) b^2 (4) $\sin x \cos x - b$ 21 _____

22 What is the solution set of the inequality $x^2 - x - 6 < 0$?

(1) $-2 < x < 3$ (3) $x < -2$ or $x > 3$
(2) $-3 < x < 2$ (4) $x < -3$ or $x > 2$ 22 _____

23 In a circle, a central angle whose measure is $\dfrac{\pi}{2}$ radians
intercepts an arc whose length is $\dfrac{3\pi}{2}$ centimeters. How
many centimeters are in the radius of the circle?

(1) 1   (3) 3
(2) 2   (4) 4   23 _____

24 The graph of the equation $y = 4^{-x}$ lies in Quadrants?

(1) I and II   (3) III and IV
(2) II and III   (4) I and IV   24 _____

25 On a standardized test with a normal distribution, the mean
is 88. If the standard deviation is 4, the percentage of grades
that would be expected to lie between 80 and 96 is closest to

(1) 5   (3) 68
(2) 34   (4) 95   25 _____

26 What is the inverse of the function $y - 2 = 7x$?

(1) $y = \dfrac{2 - x}{7}$ (3) $y = 7x - 2$

(2) $y = \dfrac{2x}{7}$ (4) $y = \dfrac{x - 2}{7}$ 26 _____

27 What is the maximum number of distinct triangles that can
be formed if $m\angle A = 30$, $b = 8$, and $a = 5$?

(1) 1   (3) 3
(2) 2   (4) 0   27 _____

28 The fraction $\dfrac{b+\dfrac{b}{a}}{a-\dfrac{1}{a}}$ is equivalent to

(1) b

(3) $\dfrac{2ab}{a^2-1}$

(2) $\dfrac{b}{a-1}$

(4) $\dfrac{a-1}{b}$

28 _____

29 As angle θ increases from π radians to 2π radians, the cosine of θ

(1) increases throughout the interval
(2) decreases throughout the interval
(3) increases, then decreases
(4) decreases, then increases

29 _____

30 The graph of which equation forms an ellipse?

(1) $x^2 - y^2 = 9$
(2) $2x^2 + 2y^2 = 8$

(3) $2x^2 + y^2 = 8$
(4) $xy = -8$

30 _____

31 Which term describes the roots of the equation $2x^2 + 3x - 1 = 0$?

(1) rational
(2) irrational

(3) equal
(4) imaginary

31 _____

32 What is the value of the expression $2x^{-\frac{1}{3}}$ when $x = 8$?

(1) 1

(3) $\dfrac{1}{2}$

(2) 2

(4) $\dfrac{1}{4}$

32 _____

33 What are the sum (S) and product (P) of the roots of the equation $3x^2 - 7x + 12 = 0$?

(1) $S = 7, P = 12$

(2) $S = \dfrac{7}{3}, P = -4$

(3) $S = \dfrac{7}{3}, P = 4$

(4) $S = -\dfrac{7}{3}, P = -4$

33 _____

34 In simplest form, what is the third term in the expansion of $(x - 2y)^6$?

(1) $-80x^3y^3$ (3) $-60x^4y^2$

(2) $80x^3y^3$ (4) $60x^4y^2$

34 _____

35 In the set of real numbers, what is the domain of $f(x) = \sqrt{x+5}$?

(1) $x \geq -5$ (3) $x > -5$

(2) $x \leq -5$ (4) $x \geq 0$

35 _____

PART II

Answer four questions from this part. Clearly indicate the necessary steps, including appropriate formula substitutions, diagrams, graphs, charts, etc. Calculations that may be obtained by mental arithmetic or the calculator do not need to be shown. [40]

36 In the accompanying diagram of circle O, tangent $\overline{PA}$, secant $\overline{PGFB}$, diameter $\overline{AOEB}$, and chord $\overline{CEFD}$ are drawn; $m\widehat{CA} = 70$; $m\widehat{DG} = 90$; and $m\angle CEA = 40$.

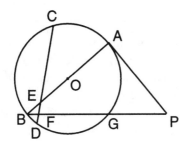

Find:

a $m\widehat{CB}$ [2]

b $m\widehat{BD}$ [2]

c $m\angle APB$ [2]

d $m\angle PAB$ [2]

e $m\angle ABG$ [2]

37 *a* On the same set of axes, sketch and label the graphs of the equations $y = \sin 2x$ and $y = 3\cos x$ in the interval $-\pi \le x \le \pi$. [8]

b Based on the graphs drawn in part *a*, find all values of x in the interval $-\pi \le x \le \pi$ that satisfy the equation $\sin 2x = 3\cos x$. [2]

38 *a* Express the roots of the equation $9x^2 = 2(3x - 1)$ in simplest $a + bi$ form.   [5]

   b Solve for *x*:

   $$\frac{12}{x^2 - 16} - \frac{24}{x - 4} = 3 \quad [5]$$

39 Find all values of θ in the interval $0 \leq \theta < 360°$ that satisfy the equation $\sin \theta = 2 + 3 \cos 2\theta$. Express your answer to the *nearest ten minutes* or *nearest tenth of a degree.*   [10]

40 *a* Peter (*P*) and Jamie (*J*) have computer factories that are 132 miles apart. They both ship their completed computer parts to Diane (*D*). Diane is 72 miles from Peter and 84 miles from Jamie. Using points *D*, *J*, and *P* to form a triangle, find m$\angle PDJ$ to the *nearest ten minutes* or *nearest tenth of a degree.*   [6]

   b If $\log 2 = a$ and $\log 13 = b$, express in terms of *a* and *b*:

   (1) $\log 26$ [1]

   (2) $\log \dfrac{8}{\sqrt{13}}$ [3]

41 *a* When $\sin x = -\dfrac{8}{17}$ and *x* lies in Quadrant III and $\cos y = -\dfrac{4}{5}$ and *y* lies in Quadrant II, what is $\cos(x - y)$?   [5]

   b For all values of *x* for which the expressions are defined, prove the following is an identity:

   $$\frac{\cos 2x}{\sin x} + \sin x = \csc x - \sin x \quad [5]$$

42 *a* Given: $J = -2 + 5i$ and $K = 3 + 2i$

    (1) On graph paper, plot and label *J* and *K*.
    [2]

    (2) On the same set of axes, plot the sum of *J* and *K* and label it *L*. [1]

    (3) On the same set of axes, plot the image of *L* after a counterclockwise rotation of 270° and label it *L′*. [2]

    (4) Express *L′* as a complex number. [1]

  b The circle in the accompanying diagram is divided into six regions of equal area and has a spinner. The regions are labeled 1, 3, 6, 9, 12, and 15. If the spinner is spun five times, what is the probability that it will land in an even-numbered region *at most* two times? [4]

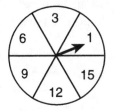

Examination
January 2002

Sequential Math Course III

PART I

Answer **30** questions from this part. Each correct answer will receive **2** credits. No partial credit will be allowed. Write your answers in the spaces provided. Where applicable, answers may be left in terms of π or in radical form.   [60]

1 Express $\dfrac{10\pi}{3}$ radians in degree measure.

1 _____

2 In the accompanying diagram, $\overline{BA}$ is a diameter and $m\overset{\frown}{BC} =$ 50. Find $m\angle CBA$.

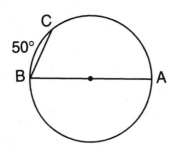

2 _____

3 Find the value of $27^{\frac{4}{3}}$.

3 _____

4 Solve for x: $\dfrac{5}{4x} - \dfrac{6}{3x} = \dfrac{1}{12}$

4 _____

5 In $\triangle ABC$, $m\angle A = 35$, $m\angle C = 60$, and $AC = 12$ meters. Find the length of $\overline{BC}$ to the *nearest meter.*

5 _____

6 In the accompanying diagram of a circle, chords $\overline{AB}$ and $\overline{CD}$ intersect at E, $CE = 5$, $CD = 13$, and $AE = 4$. Find the length of $\overline{BE}$.

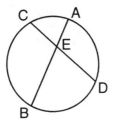

6 _____

7 In a circle with a radius of 4 centimeters, what is the number of radians in a central angle that intercepts an arc of 24 centimeters?

7 _____

8 If x varies inversely as y, and $x = 8$ when $y = 3$, find the value of x when $y = 6$.

8 _____

9 Express in simplest form: $\dfrac{\dfrac{1}{a}}{\dfrac{1}{a} - \dfrac{1}{b}}$

9 _____

10 Evaluate: $\displaystyle\sum_{k=3}^{6} k^2$

10 _____

11 If $f(x) = x^2 + 3$ and $g(x) = x - 2$, find $(f \circ g)(2)$.

11 _____

12 Express in simplest form: $\sqrt{48} - 5\sqrt{27} + 2\sqrt{75}$

12 _____

Directions (13–35): For *each* question chosen, write in the spaces provided the *numeral* preceding the word or expression that best completes the statement or answers the question.

13 Which graph does *not* represent a function?

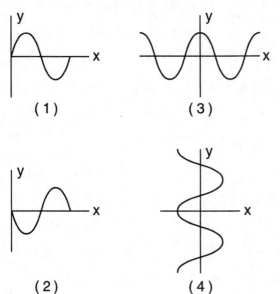

(1)

(3)

(2)

(4)

13 _____

14 The amplitude of the graph of the equation $y = \frac{1}{2}\sin 3x$ is

(1) $\frac{2\pi}{3}$

(3) $\frac{1}{2}$

(2) 6π

(4) $\frac{2}{3}$

14 _____

15 If $7.289 \times 10^n = 0.007289$, what is the value of n?

(1) –2

(3) 3

(2) 2

(4) –3

15 _____

16 The value of $\sin 170° \cos 20° - \cos 170° \sin 20°$ is

(1) $\frac{1}{2}$

(3) $\frac{\sqrt{3}}{2}$

(2) $-\frac{1}{2}$

(4) $-\frac{\sqrt{3}}{2}$

16 _____

17 If $\sin A = -\dfrac{5}{13}$ and $\cos A > 0$, angle A terminates in

Quadrant

(1) I   (3) III
(2) II   (4) IV   17 _____

18 Which expression is equivalent to $\dfrac{x^3}{x+3} - \dfrac{9x}{x+3}$?

(1) $\dfrac{-9x}{x+3}$ (3) $\dfrac{x^2}{x+3}$

(2) $\dfrac{x}{x+3}$ (4) $x(x-3)$ 18 _____

19 What is the best approximation for the area of a triangle with
consecutive sides of 4 and 5 and an included angle of 59°?

(1) 5.0   (3) 10.0
(2) 8.6   (4) 17.1   19 _____

20 If $A = \pi r^2$, $\log A$ equals

(1) $2\log \pi + \log r$ (3) $2\log \pi + 2\log r$
(2) $\log \pi + 2\log r$ (4) $2\pi \log r$ 20 _____

21 The value of $\tan(\text{Arc}\sin 1)$ is

(1) 1   (3) 90
(2) −1   (4) undefined   21 _____

22 The solution set of the equation $3^{x^2+x} = 9$ is

(1) $\{1\}$ (3) $\{-2,1\}$
(2) $\{-2\}$ (4) $\{-1,2\}$ 22 _____

23 The solution set of the equation $|2x-1| + 4 = 8$ is

(1) $\left\{\dfrac{5}{2}\right\}$ (3) $\left\{-\dfrac{3}{2}\right\}$

(2) $\left\{\dfrac{5}{2}, -\dfrac{3}{2}\right\}$ (4) $\{\ \ \}$ 23 _____

24 What is the image of point $(-1,3)$ after a reflection in the line
    $x = 2$?

    (1) $(5,3)$ (3) $(-1,1)$
    (2) $(3,3)$ (4) $(-1,-3)$ 24 _____

25 For all values of θ for which the expression is defined, $\dfrac{\sec\theta}{\csc\theta}$
    is equivalent to

    (1) $\sin\theta$ (3) $\tan\theta$
    (2) $\cos\theta$ (4) $\cot\theta$ 25 _____

26 In the accompanying diagram of a unit circle, the ordered
    pair (x,y) represents the point where the terminal side of θ
    intersects the unit circle.

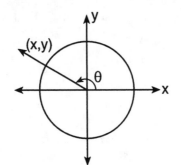

    If $\theta = 150°$, what is the value of x?

    (1) 1 (3) $-\dfrac{1}{2}$

    (2) $-\dfrac{\sqrt{3}}{2}$ (4) $-\dfrac{\sqrt{2}}{2}$ 26 _____

27 Which equation has roots $3 - i$ and $3 + i$?

    (1) $x^2 - 10x + 6 = 0$ (3) $x^2 + 6x + 10 = 0$
    (2) $x^2 + 10x - 6 = 0$ (4) $x^2 - 6x + 10 = 0$ 27 _____

28 What is the domain of the function $f(x) = \dfrac{4}{\sqrt{x+5}}$ over the set
    of real numbers?

    (1) $\{x \mid x > -5\}$ (3) $\{x \mid x \geq -5\}$
    (2) $\{x \mid x < -5\}$ (4) $\{x \mid x = -5\}$ 28 _____

29 What is one possible value of θ in the equation $\cot\theta = \cos\theta$?

(1) 0°   (3) 90°
(2) 45°   (4) 180°   29 _____

30 The graph of the equation $4x^2 + 3y = 8$ forms

(1) a straight line   (3) a hyperbola
(2) an ellipse   (4) a parabola   30 _____

31 If $m\angle A = 28° \, 10'$, $a = 20$, and $b = 25$, what is the maximum number of distinct triangles that can be constructed?

(1) 1   (3) 3
(2) 2   (4) 0   31 _____

32 The solution set for the inequality $x^2 + 4x - 5 \geq 0$ is

(1) $-5 \leq x \leq 1$ (3) $x \leq -5$ or $x \geq 1$
(2) $x \leq -1$ or $x \geq 5$ (4) $-1 \leq x \leq 5$ 32 _____

33 To the *nearest degree*, what is the measure of the largest angle in a triangle with sides measuring 10, 12, and 18 centimeters?

(1) 109   (3) 71
(2) 81   (4) 32   33 _____

34 The roots of the equation $x^2 - 6x + 7 = 0$ are

(1) imaginary
(2) real and irrational
(3) real, rational, and unequal
(4) real, rational, and equal   34 _____

35 The expression $(i^3 - 1)(i^3 + 1)$ is equivalent to

(1) -2 (3) $2i + 1$
(2) $2i - 1$ (4) $-2i$ 35 _____

PART II

Answer four questions from this part. Clearly indicate the necessary steps, including appropriate formula substitutions, diagrams, graphs, charts, etc. Calculations that may be obtained by mental arithmetic or the calculator do not need to be shown. [40]

36 Find, to the *nearest ten minutes* or *nearest tenth of a degree*, all values of x in the interval $0° \leq x < 360°$ that satisfy the equation $6\sin x + 3 = 2\csc x$. [10]

37 *a* On the same set of axes, sketch and label the graphs of the equations $y = -3\cos x$ and $y = \tan x$ in the interval $-\pi \leq x \leq \pi$. [8]

   b Using the same graph sketched in part *a*, find the number of values of x in the interval $-\pi \leq x \leq \pi$ that satisfy the equation $-3\cos x = \tan x$. [2]

38 *a* On the same set of axes, sketch and label the graphs of the equations below:

   (1) $xy = 9$ in the interval $-9 \leq x \leq 9$ [3]
   (2) $y = 3^x$ in the interval $-2 \leq x \leq 2$ [3]

   b On the same set of axes, sketch the reflection of $y = 3^x$ in the *x*-axis in the interval $-2 \leq x \leq 2$. Label the reflection *b*. [2]

   c Write an equation of the graph sketched in part *b*. [2]

39 In the accompanying diagram of circle O, tangent $\overline{AB}$ and chord $\overline{BC}$ are drawn, secant $\overline{ACD}$ intersects diameter $\overline{EB}$ at F, m$\overparen{BD}$ = 160, and m$\overparen{BC}$ = 80.

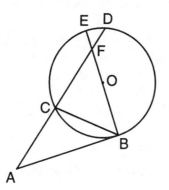

Find:

a m∠A [2]

b m∠ABE [2]

c m∠ABC [2]

d m∠EFC [2]

e m∠ACB [2]

40 a The playground at a day-care center has a triangular-shaped sandbox. Two of the sides measure 20 feet and 14.5 feet and form an included angle of 45°. Find the length of the third side of the sandbox to the *nearest tenth of a foot.*   [6]

b Given: $y = 4.1^{x}$

Find x, to the *nearest tenth,* when $y = 26$.   [4]

41 *a* Christina participated in 20 basketball games this season. The scorekeeper recorded the number of "shots" she attempted in each game. The table below shows the number of shots she attempted in the number of games she played.

Shots Attempted	Number of Games
10	4
13	3
17	5
23	6
33	2

   (1) Find the mean number of shots that Christina attempted. [1]
   (2) Find the standard deviation of the shots attempted to the *nearest tenth.* [3]
   (3) What is the total number of games in which the number of shots attempted fell outside one standard deviation of the mean? [1]

   b The probability that bus *A* will arrive on time is $\dfrac{5}{6}$. Yolanda takes this bus on 4 consecutive days. Find the probability that this bus will arrive on time:

   (1) all 4 days [2]
   (2) *at least* 3 days [3]

42 *a* Solve for *x* and express your answer in simplest *a* + *bi* form:

$$\frac{x^2}{4} = x - 2 \quad [5]$$

   b Prove the following identity:

$$\frac{\sin\theta}{\sin^2\theta + \cos 2\theta} = \frac{\sec\theta}{\cot\theta} \quad [5]$$

Answers to Regents Examinations

JUNE 2001

1. $\dfrac{5\pi}{3}$

2. 5

3. 40

4. -3

5. 24

6. II

7. $\dfrac{1}{2}$

8. 7

9. 4.55

10. $-4, \dfrac{2}{3}$

11. 80

12. 14

13. $\dfrac{32}{81}$

14. 70

15. 8

16. (2)

17. (1)

18. (2)

19. (2)

20. (4)

21. (3)

22. (1)

23. (3)

24. (1)

25. (4)

26. (4)

27. (2)

28. (2)

29. (1)

30. (3)

31. (2)

32. (1)

33. (3)

34. (4)

35. (1)

36. a 110
    b 10
    c 50
    d 90
    e 40

37. $b \dfrac{-\pi}{2}, \dfrac{\pi}{2}$

38. $a \dfrac{1}{3} \pm \dfrac{1}{3} I$
    b $-6, -2$

39. 54.6°, 123.6°, 270° or
    56°30′, 123°30′, 270°

40. a 115.4° or 115°20′
    b (1) $a + b$
    (2) $3a - \dfrac{1}{2}b$

41. $a \dfrac{36}{85}$

42. a (4) $7 - I$
    $b \dfrac{192}{243}$

JANUARY 2002

1. 600
2. 65
3. 81
4. −9
5. 7
6. 10
7. 6
8. 4
9. $\dfrac{b-b}{a}$
10. 86
11. 3
12. $-\sqrt{3}$
13. 4

14. 3
15. 4
16. 1
17. 4
18. 4
19. 2
20. 2
21. 4
22. 3
23. 2
24. 1
25. 3
26. 2

27. 4
28. 1
29. 3
30. 4
31. 2
32. 3
33. 1
34. 2
35. 1
36. 22.3°, 157.7°, 241.5° or 22°20′, 157°40′, 241°30′, 298°30′
37. b 2
38. c $y = -3^x$

39. a 40
b 90
c 40
d 130
e 100
40. a 14.1
b 2.3
41. a (1) 18.4
   (2) 6.8
   (3) 6
b (1) $\dfrac{625}{1{,}296}$
   (2) $\dfrac{1{,}125}{1{,}296}$

42. a $2 \pm 2i$

WHAT IS THE MATHEMATICS B REGENTS EXAMINATION?

The last administration of the Course III Regents Examination will be January 2004. If you need to take an advanced Regents mathematics examination after that time, you will take the Mathematics B Regents Examination. This examination covers:

- All topics from Course III except proving original trigonometric identities.
- Geometric and coordinate proofs that are included in Course II. Logic proofs are *not* required.
- Calculation of standard deviation and regression equations using a graphing calculator.
- Application of mathematical and graphing calculator skills to solving problems that are constructed in real-world settings.

HOW THE MATHEMATICS B REGENTS EXAM BREAKS DOWN

The Mathematics B Regents Exam is a 3-hour test that consists of four parts with a total of 34 questions. An actual Mathematics B Regents Examination is included at the end of this section. The accompanying table shows the point breakdown of the test.

Part	Number of Questions	Point Value	Total Points
I	20 multiple choice	2 each	$20 \times 2 = 40$
II	6	2 each	$6 \times 2 = 12$
III	6	4 each	$6 \times 4 = 24$
IV	2	6 each	$2 \times 6 = 12$
	Test = 34 questions		Test = 88 points

- Part I consists of 20 standard multiple-choice questions, each with four possible answer choices labeled (1), (2), (3), and (4).
- Questions in Parts II, III, and IV require that you show or explain how you arrived at each answer by indicating the necessary steps you take, including appropriate formula substitutions, diagrams, graphs, and charts.
- All of the questions in each of the four parts of the test must be answered.
- The answers and the work for the questions in Parts II, III, and IV must be written directly in the question booklet in the space provided underneath the questions. All work should be written in pen, except for graphs, which should be done in pencil. If you need graph paper, it will be provided in the question booklet.
- The maximum total raw score for the Mathematics B Regents Exam is 88 points. After the raw scores for the four parts of the test are added together, a conversion table provided by the New York State Education Department is used to convert you raw score into a final test score that falls within the usual 0 to 100 scale.

GRAPHING CALCULATOR SKILLS NEEDED FOR THE MATHEMATICS B REGENTS EXAMINATION

Unlike the Course III Regents Examination, the Mathematics B Regents Exam requires the use of a graphing calculator. Because of its popularity, availability, and ease of use, the Texas Instruments TI-83 graphing calculator will be used as the "reference" calculator in the following discussion. If you are using a different graphing calculator, you may need to make minor adjustments in the calculator instructions so that they will work for you model. If necessary, consult the manual that came with your calculator.

1. Performing Calculations

Routine arithmetic calculations are performed in the "home screen." Enter the home screen by pressing [2nd] [MODE]. After you enter an arithmetic expression, press [ENTER] to see the answer. The result will be displayed at the end of the next line, as shown in the accompanying screen shot. Notice that the calculator key for exponentiation is [^].

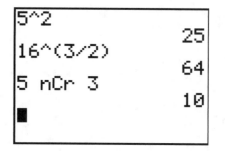

Unlike a scientific calculator, the TI-83 does not have special keys for evaluating combinations and permutations. Instead, you will need to access the calculator's math library of special functions. To evaluate $_5C_3$ (or $_5P_3$) using a TI-83 calculator:

- Enter 5.
- Press MATH ▷ ▷ ▷ 3 to select $_nC_r$ (or use option 2 to select $_nP_r$) from the MATH PRB menu.
- Enter 3 and press ENTER. The result is **10**.

2. Graphing Functions and Setting Windows

Before you can graph the function $y = x^2 - 6x + 8$, you must enter it in the Y= editor. Variable x is entered by pressing $\boxed{x,T,\theta,n}$. Notice that it is possible to enter more than one equation.

```
Plot1 Plot2 Plot3
\Y1■X^2-6X+8
\Y2=■
\Y3=
\Y4=
\Y5=
\Y6=
\Y7=
```

Set an appropriate viewing window by pressing ZOOM and then selecting option 4 for a decimal window or option 6 for a standard window. In a basic decimal window, the axes are scaled so that $-4.7 \le x \le 4.7$ and $-3.1 \le y \le 3.1$, allowing the screen cursor to move in friendly steps of 0.1. In a decimal window, the coordinates axes are scaled so that there are 10 tic marks on either side of the origin on both axes.

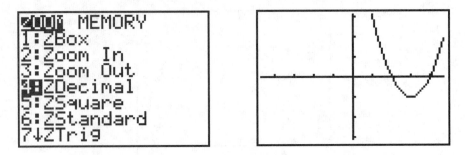

If you find that the graph does not fit in a basic decimal window, you can multiply the screen dimensions by a whole number by pressing ⎡WINDOW⎤ and then multiplying the values of Xmin and Xmax by the same whole number. The next set of screen shots shows the graph of $Y_1 = x^2 - 6x + 8$ in a friendly window in which each of the basic decimal values of Xmin and Xmax has been multiplied by 2.

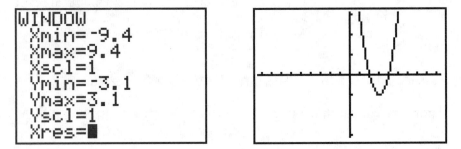

3. Creating Tables

If you want to graph $Y_1 = x^2 - 6x + 8$ using graph paper, you will need a table of values. After the equation of the parabola has been entered in the Y= menu, press ⎡2nd⎤ ⎡GRAPH⎤. A table will be displayed in which the value of x increases in steps of 1.

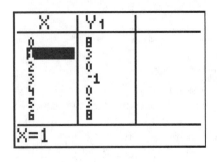

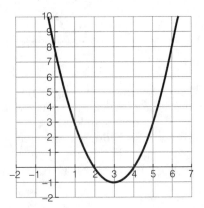

To see table values that are not in the current window, press either the up or down cursor key. If you need x to increase in increments other than 1, press [2nd] [WINDOW]. To increase x in steps of 0.5, set $\Delta Tbl = 0.5$.

4. Finding Intersection Points

If two graphs intersect, you can find the coordinates of their point(s) of intersection using the TRACE or INTERSECT feature of your graphing calculator. To solve the system of equations $y = 3x - 5$ and $y = x - 1$ graphically, enter both equations in the Y= menu. Then graph these equations in a decimal window.

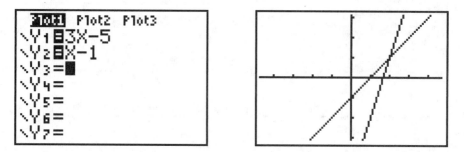

- To obtain the intersection, press [TRACE]. Then press the appropriate cursor key so that the cursor moves along one of the lines. Stop when the cursor appears to coincide with the point of intersection. The readout will be $x = 2$, $y = 1$.

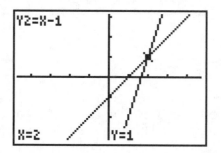

To verify that point (2,1) is the point of intersection, toggle between the two curves by pressing the up or down cursor key. The equation in the top left corner of the screen indicates the graph on which the cursor is currently located. As the cursor position changes from one graph to the other, the equation in the top left corner of the screen changes, but the coordinates at the bottom of the screen remain set at $x = 2$ and $y = 1$. This confirms that point (2,1) is common to both lines and is, therefore, the point of intersection.

- You can also obtain the point of intersection by pressing $\boxed{\text{2nd}}$ $\boxed{\text{TRACE}}$ and then selecting option **5:intersect** from the CALCULATE menu. Move the cursor to a point on the first curve that is close to the point of intersection, and press $\boxed{\text{ENTER}}$. Then move the cursor on the second curve to a point that is close to the point of intersection and press $\boxed{\text{ENTER}}$ two times. The coordinates of the point of intersection will appear at the bottom of the screen.

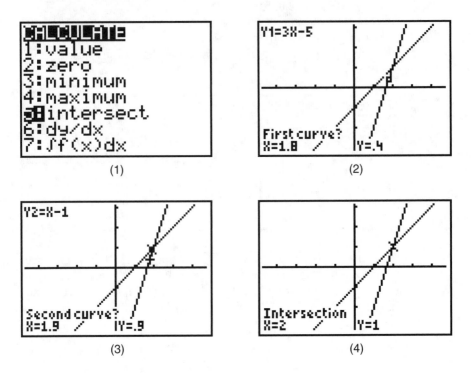

(1) (2) (3) (4)

5. Finding x-Intercepts

To find an x-intercept or zero of the graph of $y = x^2 - 4x + 2$, select **2: zero** from the CALC menu. Move the cursor to a point on the graph that is slightly to the left of the first x-intercept, and press $\boxed{\text{ENTER}}$. Then move the cursor to a point on the graph that is slightly to the right of the same x-intercept. Press $\boxed{\text{ENTER}}$ two times to display the coordinates of that x-intercept.

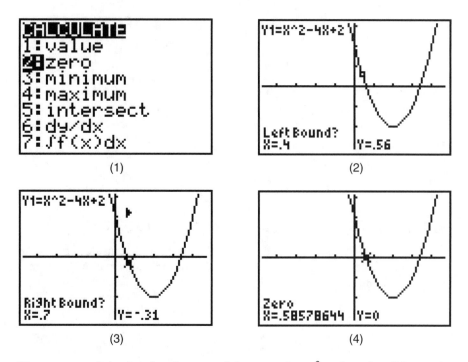

(1)

(2)

(3)

(4)

Hence, one of the irrational roots of the equation $x^2 - 4x + 2 = 0$ is, to the *nearest hundredth*, 0.59.

6. Graphing Logarithmic and Exponential Functions

The TI-83 can graph a logarithmic function whose base in 10. To graph a logarithmic function that has a base other than 10, use the change of base formula:

$$\log_b x = \frac{\log x}{\log b}.$$

For example, to graph $y = \log_3 x$ using your calculator, graph $Y_1 = \dfrac{\log(x)}{\log(3)}$.

7. Solving Complicated Equations

Any equation of the form $f(x) = g(x)$ that has real roots can be solved with a graphing calculator by graphing $Y_1 = f(x)$ and $Y_2 = g(x)$ on the same set of axes and then finding the x-value(s) of each point of intersection. Consider the equation $A = 10(0.8)^t$, where A represents the number of milligrams of a 10-milligram dose of a drug that remain in the body after t hours. To approximate how much time has passed when half of the drug dose is left in the body, solve the exponential equation $5 = 10(0.8)^t$ either by using logarithms or graphically.

- *Solution 1: Solve using logarithms*

  If $5 = 10(0.8)^t$, then $\dfrac{5}{10} = (0.8)^t$. Hence, $\log 0.5 = t \log 0.8$, so

  $$t = \frac{\log 0.5}{\log 0.8}$$

  $$\approx \frac{-0.30103}{-0.09691}$$

  $$\approx \mathbf{3.1}$$

- *Solution 2: Solve graphically*

  Since $5 = 10(0.8)^t$, graph $Y_1 = 5$ and $Y_2 = 10(0.8)^t$ in an appropriate window, such as $[0, 9.4] \times [0, 6.2]$. Use the intersect feature to obtain $x \approx 3.1$ hours.

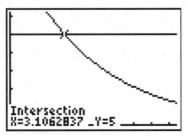

8. Graphing Trigonometric Functions

To graph a trigonometric function such as $y = 3 \sin 2x$, enter the function in the Y= menu and then press ZOOM 7. If the angular mode is set to radians, the graph will be displayed in the interval $-2\pi < x < 2\pi$ using the following preset values:

$$X\min = -\left(\frac{47}{24}\right)\pi, \qquad X\max = \left(\frac{47}{24}\right)\pi, \qquad X\mathrm{scl} = \frac{\pi}{2}, \qquad Y\mathrm{scl} = 1.$$

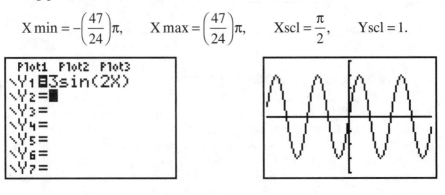

318

9. Calculating Standard Deviation

Test Score (x_i)	Frequency (f)
80	5
85	7
90	9
95	4

To calculate the standard deviation of the data in the accompanying table, press [STAT] [ENTER]. Enter the set of x-data values (scores) in list L1 and the corresponding frequencies in list L2.

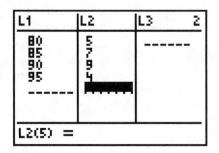

Press [STAT] [▷] [ENTER]. On the display line that begins "1-Var Stats," press [2nd] [1] [,] [2nd] [1].

```
1-Var Stats L₁,L
2█
```

Press ENTER to obtain the standard deviation, $\sigma_x \approx 4.92$.

```
1-Var Stats
 x̄=87.4
 Σx=2185
 Σx²=191575
 Sx=5.024937811
 σx=4.92341345
↓n=25
■
```

10. Using the Normal Cumulative Distribution Function (CDF)

The **normalcdf** function of your graphing calculator can be used to find the percent of scores that lie between two given scores of a normal distribution for which the mean and the standard deviation are known. For example, Mr. Yee has 184 students in his mathematics classes. Assume that the scores on the final examination are normally distributed with a mean of 72.3 and a standard deviation of 8.9. To determine the number of students that can be expected to receive scores between 82 and 90, use your graphing calculator to find the area of the normal curve that lies between the lower bound of 82 and the upper bound of 90.

- Press 2nd [VARS] 1 to select the normal cumulative density function.
- Enter the lower bound, the upper bound, the mean, the standard deviation, and the right parenthesis. Separate numerical values with commas. The display should show the following: normalcdf(82, 90, 72.3, 8.9).
- Press ENTER to get .1145178018. This value represents the percent of scores between 82 and 90. Since there are 184 students in the class, you can expect $184 \times .1145 \approx 21$ students to receive scores between 82 and 90.

11. Fitting a Regression Line to Data

If corresponding values of two sets of observed data are plotted as ordered pairs, the line or curve (exponential, power, or logarithmic) that best fits the scatter plot of data points can be obtained using the **regression** feature of a graphing calculator.

Minutes Studied (x)	15	40	45	60	70	75	90
Test Grade (y)	50	67	75	75	73	89	93

If the *x*-values in the accompanying table have already been stored in list L1 and the *y*-values stored in list L2, the regression line or line of best fit can be calculated.

- Press $\boxed{\text{STAT}}$ $\boxed{\triangleright}$ $\boxed{4}$ to choose the **LinReg**(ax + b) option.
- Then press $\boxed{\text{ENTER}}$ to show a summary of the regression statistics where *a* is the slope of the regression line, *b* is the *y*-intercept, and *r* is the coefficient of linear correlation.
- Substitute the rounded-off values of *a* and *b* into the regression equation *y* = *ax* + *b* to get **y = 0.53x + 44.70**.
- Observe that, since the coefficient of linear correlation, *r*, is close to 1, the line is a good fit to the data.

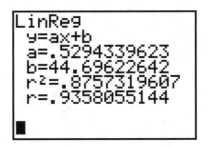

To store the regression equation as Y1, immediately after selecting the linear regression option press:

$$\boxed{\text{VARS}} \quad \boxed{\triangleright} \quad \boxed{\text{ENTER}} \quad \boxed{\text{ENTER}}.$$

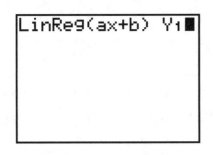

Then press $\boxed{\text{Y=}}$ to see the regression equation.

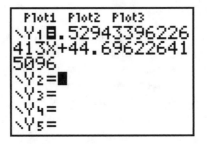

To predict the value of y when $x = 80$ minutes, substitute 80 for x in the regression equation:

$$y = 0.53x + 44.70 = 0.53 \times 80 + 44.7 \approx 87.$$

If the regression equation has been stored as Y1, you can also obtain the predicted y-value by using the table feature of your calculator:

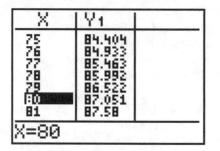

12. Fitting a Curve to Data

Data are not necessarily linearly related. For example, data that describe population growth or some analogous situation typically follow an exponential pattern. The regression menu allows you to choose other types of regression models. To fit an exponential function $y = ab^x$ to stored data, choose menu option **0:ExpReg**. A logarithmic function $y = a\ln x + b$ can be fitted to stored data by selecting menu option **9:LnReg**.

For example, the data in the accompanying table show the growth of cellular phone subscriptions in the United States from 1993 to 1999.

Year	Subscriptions (millions)
1993	16.0
1995	33.79
1996	44.04
1997	55.31
1999	86.05

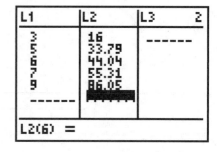

To find an exponential curve of the form $y = ab^x$ that fits the data, let x represent the year number and y represent the number of cellular phone subscribers. Enter the year numbers in list L1, where $x = 0$ is the year 1990, and store the corresponding numbers of cellular phone subscriptions in list L2 as shown in the accompanying figure. Then press

to find that the exponential regression equation, as seen from the figure below, is approximately $y = 7.746(1.319)^x$.

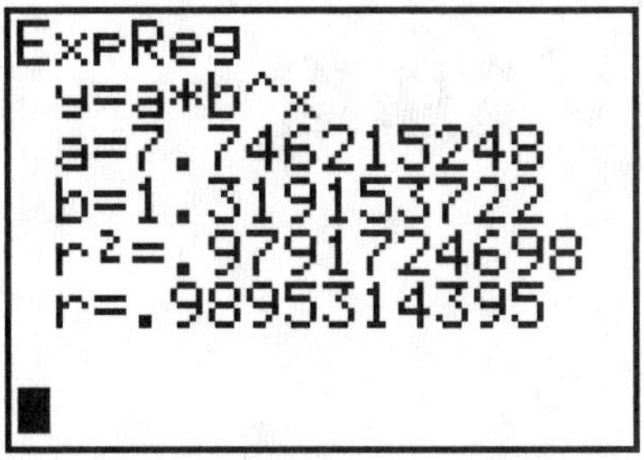

```
ExpReg
 y=a*b^x
 a=7.746215248
 b=1.319153722
 r²=.9791724698
 r=.9895314395
■
```

Examination
January 2002
Math B

FORMULAS

Area of Triangle

$$K = \frac{1}{2}ab \sin C$$

Function of the Sum of Two Angles

$\sin(A + B) = \sin A \cos B + \cos A \sin B$

$\cos(A + B) = \cos A \cos B - \sin A \sin B$

Function of the Difference of Two Angles

$\sin(A - B) = \sin A \cos B - \cos A \sin B$

$\cos(A - B) = \cos A \cos B + \sin A \sin B$

Law of Sines

$$\frac{a}{\sin A} = \frac{b}{\sin B} = \frac{c}{\sin C}$$

Law of Cosines

$$a^2 = b^2 + c^2 - 2bc \cos A$$

Functions of the Double Angle

$\sin 2A = 2 \sin A \cos A$

$\cos 2A = \cos^2 A - \sin^2 A$

$\cos 2A = 2 \cos^2 A - 1$

$\cos 2A = 1 - 2 \sin^2 A$

Functions of the Half Angle

$$\sin \frac{1}{2}A = \pm\sqrt{\frac{1-\cos A}{2}}$$

$$\cos \frac{1}{2}A = \pm\sqrt{\frac{1+\cos A}{2}}$$

Normal Curve
Standard Deviation

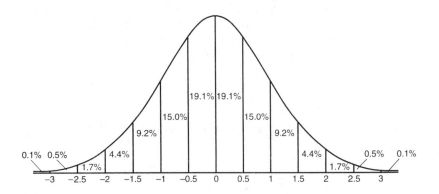

PART I

Answer all questions in this part. Each correct answer will receive 2 credits. No partial credit will be allowed. Record your answers in the spaces provided. [40]

1 The roots of a quadratic equation are real, rational, and equal when the discriminant is

(1) −2   (3) 0

(2) 2   (4) 4   1 ____

2 Chad had a garden that was in the shape of a rectangle. Its length was twice its width. He decided to make a new garden that was 2 feet longer and 2 feet wider than this first garden. If x represents the original width of the garden, which expression represents the difference between the area of his new garden and the area of the original garden?

(1) $6x + 4$ (3) $x^2 + 3x + 2$

(2) $2x^2$ (4) 8   2 ____

3 The accompanying graph represents the value of a bond over time.

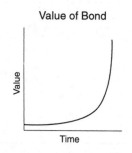

Value of Bond

Which type of function does this graph best model?

(1) trigonometric   (3) quadratic

(2) logarithmic   (4) exponential   3 ____

4 An object that weighs 2 pounds is suspended in a liquid. When the object is depressed 3 feet from its equilibrium point, it will oscillate according to the formula $x = 3 \cos (8t)$, where t is the number of seconds after the object is released. How many seconds are in the period of oscillation?

(1) $\dfrac{\pi}{4}$ (3) 3

(2) π (4) 2π 4 _____

5 If θ is an angle in standard position and its terminal side passes through the point $\left(\dfrac{1}{2}, \dfrac{\sqrt{3}}{2} \right)$ on a unit circle, a possible value of θ is

(1) 30°   (3) 120°

(2) 60°   (4) 150°   5 _____

6 The expression $\dfrac{\dfrac{a}{b} - \dfrac{b}{a}}{\dfrac{1}{a} + \dfrac{1}{b}}$ is equivalent to

(1) $a + b$ (3) ab

(2) $a - b$ (4) $\dfrac{a-b}{ab}$ 6 _____

7 If $f(x) = 5x^2$ and $g(x) = \sqrt{2x}$, what is the value of $(f \circ g)(8)$?

(1) $8\sqrt{10}$ (3) 80

(2) 16   (4) 1,280   7 _____

8 Which expression is *not* equivalent to $\log_b 36$?

(1) $6 \log_b 2$ (3) $2 \log_b 6$

(2) $\log_b 9 + \log_b 4$ (4) $\log_b 72 - \log_b 2$ 8 _____

9 If a function is defined by the equation $y = 3x + 2$, which equation defines the inverse of this function?

(1) $x = \dfrac{1}{3}y + \dfrac{1}{2}$ (3) $y = \dfrac{1}{3}x - \dfrac{2}{3}$

(2) $y = \dfrac{1}{3}x + \dfrac{1}{2}$ (4) $y = -3x - 2$ 9 _____

10 Which transformation is *not* an isometry?

(1) $r_{y=x}$ (3) $T_{3,6}$
(2) $R_{0,90°}$ (4) D_2 10 _____

11 Which relation is a function?

(1) $x = 4$ (3) $y = \sin x$
(2) $x = y^2 + 1$ (4) $x^2 + y^2 = 16$ 11 _____

12 In $\triangle ABC$, $m\angle A = 33$, $a = 12$, and $b = 15$. What is $m\angle B$ to the *nearest degree*?

(1) 41   (3) 44
(2) 43   (4) 48   12 _____

13 The accompanying diagram represents circular pond O with docks located at points A and B. From a cabin located at C, two sightings are taken that determine an angle of 30° for tangents $\overrightarrow{CA}$ and $\overrightarrow{CB}$.

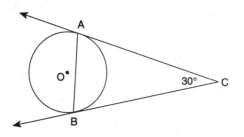

What is $m\angle CAB$?

(1) 30   (3) 75
(2) 60   (4) 150   13 _____

14 The accompanying diagram shows a section of a sound wave as displayed on an oscilloscope.

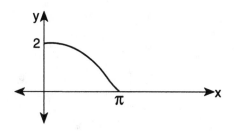

Which equation could represent this graph?

(1) $y = 2 \cos \dfrac{x}{2}$ (3) $y = \dfrac{1}{2} \cos 2x$

(2) $y = 2 \sin \dfrac{x}{2}$ (4) $y = \dfrac{1}{2} \sin \dfrac{\pi}{2} x$ 14 _____

15 Every time the pedals go through a 360° rotation on a certain bicycle, the tires rotate three times. If the tires are 24 inches in diameter, what is the minimum number of complete rotations of the pedals needed for the bicycle to travel at least 1 mile?

(1) 12   (3) 561
(2) 281   (4) 5,280   15 _____

16 Which type of symmetry does the equation $y = \cos x$ have?

(1) line symmetry with respect to the x-axis
(2) line symmetry with respect to $y = x$
(3) point symmetry with respect to the origin
(4) point symmetry with respect to $\left(\dfrac{\pi}{2}, 0 \right)$ 16 _____

17 The value of $\left(\dfrac{3^0}{27^{\frac{2}{3}}} \right)^{-1}$ is

(1) -9 (3) $-\dfrac{1}{9}$

(2) 9 (4) $\dfrac{1}{9}$ 17 _____

18 What is the domain of $h(x) = \sqrt{x^2 - 4x - 5}$?

    (1) $\{x | x \geq 1 \text{ or } x \leq -5\}$ (3) $\{x | -1 \leq x \leq 5\}$

    (2) $\{x | x \geq 5 \text{ or } x \leq -1\}$ (4) $\{x | -5 \leq x \leq 1\}$ 18 _____

19 The expression $(-1 + i)^3$ is equivalent to

    (1) $-3i$ (3) $-1 - i$

    (2) $-2 - 2i$ (4) $2 + 2i$ 19 _____

20 The revenue, $R(x)$, from selling x units of a product is represented by the equation $R(x) = 35x$, while the total cost, $C(x)$, of making x units of the product is represented by the equation $C(x) = 20x + 500$. The total profit, $P(x)$, is represented by the equation $P(x) = R(x) - C(x)$. For the values of $R(x)$ and $C(x)$ given above, what is $P(x)$?

    (1) $15x$ (3) $15x - 500$

    (2) $15x + 500$ (4) $10x + 100$ 20 _____

PART II

Answer all questions in this part. Each correct answer will receive 2 credits. Clearly indicate the necessary steps, including appropriate formula substitutions, diagrams, graphs, charts, etc. For all questions in this part, a correct numerical answer with no work shown will receive only 1 credit. [12]

21 Explain how a person can determine if a set of data represents inverse variation and give an example using a table of values.

22 Solve for x in simplest $a + bi$ form: $x^2 + 8x + 25 = 0$

23 A ball is rolling in a circular path that has a radius of 10 inches, as shown in the accompanying diagram. What distance has the ball rolled when the subtended arc is 54°? Express your answer to the *nearest hundredth of an inch.*

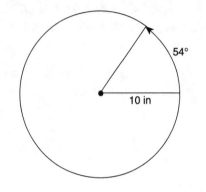

54°

10 in

24 A rectangle is said to have a golden ratio when $\dfrac{w}{h} = \dfrac{h}{w-h}$, where w represents width and h represents height. When $w = 3$, between which two consecutive integers will h lie?

25 The accompanying diagram shows the floor plan for a kitchen. The owners plan to carpet all of the kitchen except the "work space," which is represented by scalene triangle *ABC*. Find the area of this work space to the *nearest tenth of a square foot.*

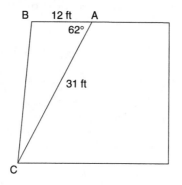

26 A set of normally distributed student test scores has a mean of 80 and a standard deviation of 4. Determine the probability that a randomly selected score will be between 74 and 82.

PART III

Answer all questions in this part. Each correct answer will receive 4 credits. Clearly indicate the necessary steps, including appropriate formula substitutions, diagrams, graphs, charts, etc. For all questions in this part, a correct numerical answer with no work shown will receive only 1 credit. [24]

27 Two straight roads, Elm Street and Pine Street, intersect creating a 40° angle, as shown in the accompanying diagram. John's house (*J*) is on Elm Street and is 3.2 miles from the point of intersection. Mary's house (*M*) is on Pine Street and is 5.6 miles from the intersection. Find, to the *nearest tenth of a mile,* the direct distance between the two houses.

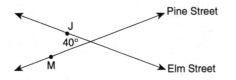

28 At the local video rental store, José rents two movies and three games for a total of $15.50. At the same time, Meg rents three movies and one game for a total of $12.05. How much money is needed to rent a combination of one game and one movie?

29 Team A and team B are playing in a league. They will play each other five times. If the probability that team A wins a game is $\frac{1}{3}$, what is the probability that team A will win *at least* three of the five games?

30 Depreciation (the decline in cash value) on a car can be determined by the formula $V = C(1 - r)^t$, where V is the value of the car after t years, C is the original cost, and r is the rate of depreciation. If a car's cost, when new, is $15,000, the rate of depreciation is 30%, and the value of the car now is $3,000, how old is the car to the *nearest tenth of a year*?

31 When a baseball is hit by a batter, the height of the ball, $h(t)$, at time t, $t \geq 0$, is determined by the equation $h(t) = -16t^2 + 64t + 4$. For which interval of time is the height of the ball greater than or equal to 52 feet?

32 *a* On the accompanying grid, graph the equation $2y = 2x^2 - 4$ in the interval $-3 \leq x \leq 3$ and label it *a*.

    b On the same grid, sketch the image of *a* under $T_{5,-2} \circ r_{x\text{-}axis}$ and label it *b*.

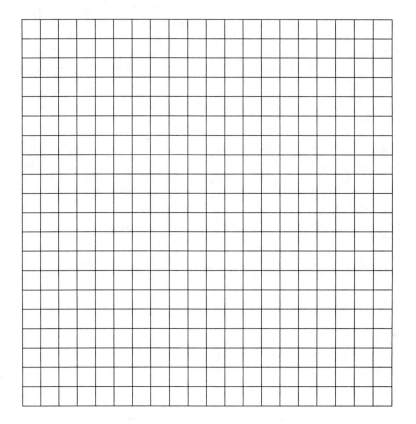

PART IV

Answer all questions in this part. Each correct answer will receive 6 credits. Clearly indicate the necessary steps, including appropriate formula substitutions, diagrams, graphs, charts, etc. For all questions in this part, a correct numerical answer with no work shown will receive only 1 credit. [12]

33 Prove that the diagonals of a parallelogram bisect each other.

34 Two different tests were designed to measure understanding of a topic. The two tests were given to ten students with the following results:

Test x	75	78	88	92	95	67	58	72	74	81
Test y	81	73	85	88	89	73	66	75	70	78

Construct a scatter plot for these scores, and then write an equation for the line of best fit (round slope and intercept to the *nearest hundredth*).

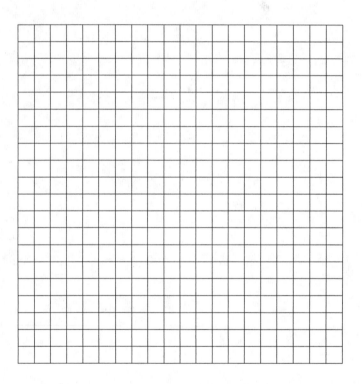

Find the correlation coefficient.

Predict the score, to the *nearest integer,* on test *y* for student who scored 87 on test *x*.

Answers to Regents Examination

JANUARY 2002

PART I

1. 3	**5.** 2	**9.** 3	**13.** 3	**17.** 2
2. 1	**6.** 2	**10.** 4	**14.** 1	**18.** 2
3. 4	**7.** 3	**11.** 3	**15.** 2	**19.** 4
4. 1	**8.** 1	**12.** 2	**16.** 4	**20.** 3

Parts II–IV You are required to show how you arrived at your answers.

PART II

21. A set of data can represent inverse variation if the product of two variables is constant. Show a correct table of values.

22. $-4 \pm 3i$

23. 9.42

24. 1 and 2, $1 < x < 2$, or $1 < 1.854 < 2$

25. 164.2

26. 0.624 or 62.4%

PART III

27. 3.8, use Law of Cosines

28. \$6.15

29. $\frac{51}{243}$

30. 4.5

31. $1 \le t \le 3$

32. $2y = 2x^2 - 4$ graphed correctly and labeled.

PART IV

33. Correct Euclidian proof, with concluding statement that the diagonals bisect each other, or correct analytic proof using coordinate geometry, with concluding statement that the diagonals bisect each other.

34. Correct scatter plot, $y = 0.62x + 29.18$, $r = 0.92$, and 83

INDEX